"The topic of protecting inventions and innovation—Intellectual Property—that Kip writes about in her book is something that affects even the more established companies like Ferragamo. Counterfeits and imitations plague the market daily. We spend a great deal of time defending against the corrosive effects on our brand.

"My father's passion for creating the Ferragamo shoe came from his years of studying human anatomy of the foot and designing the signature Ferragamo brand. Without the protection of strong patents and trademarks, there is no protecting innovation no matter what it looks like: whether it be a shoe design or a special signature clasp or logo or even a cure for cancer. For the small creatives and innovators, all the way up to the international company, what differentiates a brand is their own unique intellectual property, design, innovation. Protection of this is imperative across all sectors."

—Massimo Ferragamo, Chairman of Ferragamo U.S.A.

"Kip's story is the unfortunate story of just about all inventors in current times. Kip explains the complicated legal changes that deconstructed the patent system for small guys in simple and hilarious terms throughout her story. I recommend this book for all inventors and all Americans who want the best for our country."

—Paul Morinville, U.S. Inventor

"Patent stuff is usually written by lawyers and blowhards, and reading it is usually about as much fun as licking a chalkboard. I've read probably 20 to 25 books about the patent system, and only two have stood out as enjoyable. Kip and Scott's book is one of them. They took a rather dry subject and made it fun and engaging, which is no small feat, and got into some really thick material with more ease than I've seen."

—Will Plut, Arms Dealer in the Patent World

BLOOD
IN THE
WATER

Also by Kip Azzoni Doyle

Nonfiction

Selling Dreams
How to Make Any Product Irresistible
(with Gian Luigi Longinotti-Buitoni)

Also by Scott Burr

Fiction

Bummed Out City: A Novel
We Will Rid the World of You: A Novel

Nonfiction

Worth Defending
How Gracie Jiu-Jitsu Saved My Life
(with Richard Bresler)

Superhero Simplified
Collected, Selected, Revised and Expanded

Suspend Your Disbelief
*How to Build and Build Strength With the World's
Most Rugged Suspension Training Device*

Get a Grip
*A Practical Primer on Grip Strength and
Endurance Training... and More*

BLOOD
IN THE
WATER

AMERICA'S ASSAULT
ON INNOVATION

KIP AZZONI DOYLE
WITH
SCOTT BURR

Disclaimer

Although the authors have made every effort to ensure that the information in this book was correct at press time, and while this publication is designed to provide accurate information in regard to the subject matter covered, the authors assume no responsibility for errors, inaccuracies, omissions, or any other inconsistencies herein and hereby disclaim, to the maximum extent permitted by law, any liability to any party for any loss, damage, or disruption caused by errors or omissions, whether such errors or omissions result from negligence, accident, or any other cause.

This publication is meant as a source of valuable information for the reader; however, it is not meant as a substitute for direct expert assistance. If such level of assistance is required, the services of a competent professional should be sought. To the maximum extent permitted by law, the authors disclaim any and all liability in the event that any information, commentary, analysis, opinions, advice and/or recommendations contained in this book prove to be inaccurate, incomplete, or unreliable.

This book, in part, reflects the author's present recollections of experiences over time. Some events, places, and conversations have been recreated from memory, and interview transcripts have been edited for content and clarity. The names and details of some individuals have been changed or obscured to respect their privacy.

The views and opinions expressed in this book are solely those of Kip Azzoni Doyle, and do not necessarily reflect the views and opinions of her interview subjects, associates, or affiliates. The views and opinions expressed in interviews and/or first-person accounts are solely those of the interview subject or source, and do not necessarily reflect the views and opinions of other interview subjects, or their own associates or affiliates.

While every effort has been made to obtain permission from the owners of reproduced copyrighted material, the authors apologize for any omissions and will be pleased to incorporate missing acknowledgements in any future editions.

Kip's Dedication:

Dedicating this to my believers: Kalypso, Celeste, Sybilla, and Marco.
Yeah, that'd be my four kids, whose names I've been told sound collectively like
an exotic fruit salad.

To my superhero husband, Mark. He wields a mean red pen.

…and to Geronimo, of course.

CO-AUTHOR'S PREFACE:

CLIMB ON

When I first got involved with this project, I didn't know the first thing about patents. Or maybe, to put it more accurately, I knew what everybody "knows" about them: inventors invent things, secure patents for their inventions, and these patents keep other people from ripping those inventions (and those inventors) off. I certainly had never heard of the America Invents Act; nor did I have any idea that, in the years since that Act's passage in 2011, America has been on the fast-track to an innovation crisis.

But we'll get into all that. This preface isn't so much about the *content* of this book as it is about the nature of the book itself. As Will Plut so eloquently stated in his very generous blurb, reading most of what is written about the patent system is about as much fun as licking a chalk board—a sentiment I can wholeheartedly echo. I had plenty of catching up to do when I got involved with this book—plenty of patent history, recent and less recent, to familiarize myself with—and "licking a lot of chalk boards" is about as good a way to describe that project as any. It's a classic case of the internet's double-edged sword cutting both ways: you have access to all the information you could ever want—and more—if you only have the wherewithal to dig through it.

As you'll figure out pretty quickly in reading this book, Kip is neither your typical inventor nor your typical writer-on-the-subject-of-patents. This is not meant as a knock on either of these; however, at the same time, it must be said that Kip is a singular personality. Smart, funny, passionate, engaged, driven... I could go on, but I think you'll get the idea quickly enough when you get into the body of this book. The point being, it was clear to me when I came on-board that writing a book about the state of the patent system and writing *Kip's* book about the state of the patent system were different things, and a book that succeeded as the former and failed as the latter would be a disservice to Kip, her story, the broader subject, and the book's intended audience. Accordingly, as we got going, Kip and I spent a lot of time discussing exactly what we want-ed this book to be (at one point Kip said that she wanted it to be like a "plastic hula girl, all hippy and jiggling, on your Honda dash"), how we wanted to present its content to its readers (i.e. NOT like licking a chalk board), and where we wanted it to fall in those readers' education (we provide a lot of re-sources in this book and it is our hope that, when you're done, you'll be in-spired to go check them out). Among the concerns we had in building this book, none was more central than *accessibility of information*. After all: a book that nobody wants to read—or that, once started, nobody wants to fin-ish—is functionally equivalent to there being no book at all. If we were going to do this thing then we needed to do it well and we needed to do it right.

We needed to do it well and we needed to do it right because the truth is that the American patent system is in serious trouble. The fundamental changes that crashed in upon this foundational institution with the passage and imple-mentation of the America Invents Act have been, in a word, disastrous—on a par, one might argue, with the changes that precipitated the financial crisis of 2008. But whereas the average American adult can engage with at least the broad concept of mortgage lending from personal experience, and might then be made to understand the relationship between speculative lending, toxic assets, and the recession that resulted, the situation at the United States Patent and Trademark Office—no less threatening to our nation's health—is far more abstract to the average reader. Many, many people secure mortgages to buy homes; far fewer people secure patents. It is, almost by definition, a seemingly niche concern... only it isn't. A decimated patent system affects all of us, though its effects may be less clearly linked to their source. Illustrating and explaining this, and doing it in an effective and engaging way, was of paramount importance... and all of that got me thinking about the job I had back in college.

I went to school in Colorado, mostly to be near the mountains. I'd been bitten by the climbing bug in my teens, and I was thrilled to escape my native Ohio's sparse supply of non-architectural vertical terrain. I spent my four-year term of study exploring the climbing in and around Colorado Springs and the South Platte, including a lot of time in the world-famous and tourist-choked Garden of the Gods, and working at the school's on-campus rock climbing gym. I started out at the front desk, moved up to route-setter, then head route-setter, then co-chair of the school's climber's association. Of these jobs, I liked route-setting the best. Those of you who've spent any time doing it know that rock climbing is basically a form of physical problem solving: the available holds present both an opportunity for and a challenge to upward progress, and it is by unlocking the specific sequence of movements and orientations required that a climber completes the route. Route-setting, then, works in something like the reverse, complicating the direct line with various inclusions or exclusions, orchestrating unintuitive sequences, and otherwise endeavoring to frustrate the aspirant.

However. Route-setting at a community climbing gym requires that you take this skill and apply it across a spectrum. Again: accessibility is key. A gym full of extraordinarily challenging routes will attract those (few) with the skill and strength to engage meaningfully with the routes; everyone else will move on to other activities. As one of a cadre of route-setters I had the luxury of setting mostly routes that I and my climbing partners found interesting; as the head route-setter, I had to broaden my approach. What would a novice climber find challenging but manageable? An intermediate climber? A climber stronger and better than me? The job took on a different aspect. The puzzle wasn't just about the route itself: now it was about the people who would be climbing that route. Later, as I became more experienced as a writer, it struck me that these undertakings were not dissimilar: writing for myself, to express what I felt I had to say, and even to express it well, was only a part of the undertaking; the better part wasn't about me at all. The better part was about giving an experience to my reader: about earning their time, effort, and attention by focusing on what I was offering them in return.

All of which is a long way of saying that, in building this book, we endeavored to give you as many hand and footholds as you need, to explicate the challenging bits and make the challenges they present surmountable, and to otherwise give Kip's story, the material to which it leads, and the message contained therein the platform they deserve and the best chance to connect with as many of you as possible. Because the simple fact is that the story of America's assault on innovation needs to be heard. The problem is profound, the situation is devolving as we speak, and the stakes couldn't be higher. There is still time to set the ship aright, but we need to take action and we need to take it now. It is our hope that by the time you finish this book you'll agree, and that you'll join your voice with ours in the effort to undo the damage done by the America Invents Act, for indie inventors' sake as well as your own. The future depends on it.

Scott Burr
Cleveland, Ohio
9/26/21

INTRODUCTION

"Where are we now?
The moment you know,
you know, you know."
—David Bowie, "Where Are We Now?"

In an April 2020 article published on Reuters.com, reporter Stephanie Nebehay outlines how, in 2019, China "was the biggest source of applications for international patents in the world… pushing the United States out of the top spot it has held since the global system was set up more than 40 years ago." According to the World Intellectual Property Organization (WIPO), the international organization established to oversee the system within which the ranking occurred and within which countries share recognition of patents, China's 58,990 applications represented "a 200-fold increase in just 20 years."

Ownership of patents, Nebehay noted, is "widely seen as an important sign of a country's economic strength and industrial know-how"; the United States, she said, "had filed the most applications in the world every year since the Patent Cooperation Treaty system was set up in 1978"; now, however, "[m]ore than half of patent applications—52.4%—come from Asia."

The WIPO's head, Francis Gurry, told a news conference that China's success came "down to a very deliberate strategy on the part of Chinese leadership to advance innovation and to make the country a country whose economy operates at a higher level of value."[1] He did not, however, comment on the flip side of the coin: America's fall from the top spot, and the nearly 10-year-long assault it had levied against its own innovators which made that fall inevitable.

The American patent system has always presented something of a paradox. In fact, for a country propagating what might be described as a "Darwinian" view of market economics, the patent system presents an outright anomaly: in simple terms, patents—and the temporary monopoly they offer over a technology, product, design, et cetera—represent an absolute cessation of the competition that is thought to drive the economy and the country forward. On the other hand, as inventor and intellectual property commentator Neal Solomon points out in his 2017 article "The Disintegration of the American Patent System," published on the patent and intellectual property news site IPWatchdog.com, the patent system "provides an incentive to invest in risky technical problem-solving by giving an inventor an 'exclusive right' for a limited time."[2] The tension between these trends, Solomon continues, has "enabled the rise of the U.S. as a major industrial economy, particularly after the Civil War." And, according to Solomon, the system "worked well, as

---

[1] "In a first, China knocks U.S. from top spot in global patent race," Stephanie Nebehay, Reuters, April 7, 2020. https://www.reuters.com/article/us-usa-china-patents/in-a-first-china-knocks-u-s-from-top-spot-in-global-patent-race-idUSKBN21P1P9

[2] "The Disintegration of the American Patent System" Neal Solomon, IP Watchdog, January 26, 2017. https://www.ipwatchdog.com/2017/01/26/disintegration-american-patent-system/id=77594/

evident in the development of the largest and strongest economy in the world."

"Worked well," that is, until recently. Tellingly, and according to information from the Global Innovation Policy Center's[3] annual report and published on IPWatchdog.com,[4] the U.S. patent system's ranking in terms of the patent protection it offers innovators has been steadily falling: in 2017 it ranked 10th worldwide; by 2018 it had tumbled out of the top 10 to tie with Italy for 12th place.[5] According to the site, the overall U.S. patent ranking "decreased for the 6th consecutive year as the result of a patent climate that the [Global Innovation Policy Center] characterizes as causing 'considerable uncertainty for innovators.'"

The article goes on to note that, during this same interval, "China dramatically improved" its own ranking, "raising from a score of 4.35 (out of 8) in 2017 to a score of 5.5 (out of 8) in 2018."

The "considerable uncertainty" in the U.S. patent climate—and U.S. patent system's steady descent in the GIPC ranking—can be traced, according to the report, to one specific recent phenomenon: "how easy it has become to challenge patents in post-grant proceedings at the Patent Trial and Appeal Board (PTAB)." This, the authors note, creates "considerable uncertainty for innovators and the legal community... [which] seriously undermines the longstanding world-class innovation environment in the U.S. and threatens the nation's global competitiveness."

The GIPC's report, it should be further noted, coincides with other reports sounding the same alarm: the contemporaneous Bloomberg Innovation Index

[3] https://www.theglobalipcenter.com

[4] "U.S. Patent System Falls to 12th Place in Chamber Global IP Index for 2018" Gene Quinn, IP Watchdog, February 8, 2018. https://www.ipwatchdog.com/2018/02/08/u-s-patent-system-falls-12th-place-chamber-global-ip-index-2018/id=93494/

[5] Don't get me wrong: Italy is my motherland, and I love me some delish gelato, but seriously? Italy?

(BII) showed the U.S. falling out of the top 10 most innovative countries for the first time in the index's history; the 2021 BII report found us likewise situated, with South Korea now leading the pack.

How did this happen? How did the U.S. go from head of the class to not even cracking the top 10? The simple answer is: a misguided Congress, a Big-Tech agenda, a straw man, and a certain piece of legislation that "scorched the sky."[6] The simple answer is big money doing what big money does: peddling influence in the Halls of Power to its own short-sighted and self-serving ends. The simple answer is that the answer isn't simple at all: it's a tangled web of obscured associations and closed-door dealings, and it's going to take a flashlight and a strong stomach to get to the bottom of it. The simple answer is that *you*—the people for whom this government is of, for, and by—*deserve* an answer, and in the coming pages I intend to give you one.

Why? Because our future depends on it. Because if, as is stated by the Innovation Alliance, "cutting-edge startups and entrepreneurs—and their investors—have always been responsible for the real breakthrough inventions and technologies needed for economic growth, improved quality of life, increased productivity, and national security,"[7] then the decline in their fortunes is the decline in ours as well; because without the independent inventors we wouldn't have penicillin, duct tape, and online shopping—all the things that make modern life *so damn good*—wouldn't have the encryption technologies that keep us safe when we do that online shopping, wouldn't have the defense technologies that keep our men and women in uniform—not to mention our nation itself—safer in the face of enemy interests, wouldn't have the technologies that keep

---

[6] Read on... this will make sense later.

[7] "USIJ Report: U.S. Startup Company Formation and Venture Capital Funding Trends 2004 to 2017" Executive Summary, Innovation Alliance, July 9, 2018. https://innovation-alliance.net/patent-news/what-others-are-saying/usij-u-s-startup-company-formation-and-venture-capital-funding-trends-2004-to-2017/

us and our loved ones safer out on the roads. Because without tomorrow's independent inventors there's *no telling what we won't have...* and what others *will* have... and how they'll use it against us.

But what can We the (little) People—the consumers and the voters and the entrepreneurs and the independent inventors tinkering away in our garages late into the night—do about such a broad and entrenched and amorphous problem, a problem with Big Tech money on one side and surging global economies on the other? What can we do to right this badly listing ship? The answer is *plenty*, Young Grasshopper, but we need to do it *soon* and we need to start *now.* Time is running out, and there's a lot to learn. Don't worry, though. I know you can do it. Why? Because I did. When I started out I didn't know any of what I'm about to tell you. I never even intended to find out. Back then I was just a chick on a bike, cruising along through her own private two-wheeled daydream: a chick who tripped over her own laziness and took a short knee-and-elbow-scuffing tumble down into inspiration.

PART I

NOT THE ONLY SHARK
IN THE OCEAN

"Do your thing, lemme do my thang."
—Joe Budden, "Pump it Up"[8]

Strap in, folks, helmets on, and take a ride with me from "Concept" to "Concrete" to "WTF just happened?" This here is the trippy tale of one little inventor—me—and how she went from 1) scrawling an idea on the back of a cocktail napkin to 2) actually holding the product in her hand to 3) fighting to keep her head above water in a wicked undertow with the predatory fish getting ever closer: the trippy tale of how the U.S. patent system has been blown righteously off the rails and is now bearing down on the head of *YOU*, the little independent ("indie") inventor, with your big American dreams: the trippy tale of how we all need to get our asses in gear *right quick* before the recent changes to the U.S. patent system put the whole Red, White, and Blue ball o'

---

[8] Best lyrics evvveeerrr.

wax out of contention for good. It's a headbanger, folks, full of switchbacks and chicanes, with potholes and pitfalls and hundred-foot drops on either side. Safety is not guaranteed on the strange trip we indie inventors take to bring forth and defend our inventions, and I highly recommend that you talk to your doctor first, 'cuz this shit ain't for the weak of heart.

Sound dreamy? Peachy keen? Then climb aboard and hang on.

1. YOU DON'T HAVE TO BE A GENUIS[9]

Let's be real clear: inventing something either comes from being super smart and using your multiple degrees to cure cancer or COVID or some terrible, incurable childhood disease… or it comes from being the laziest person on earth. For me, it happened out on the road. I'm cruising along on my motorcycle[10] when I look down and see that I'm low on fuel. I roll up to the pump at the next station and I pull off my helmet. As I suck in that first glorious, intoxicating whiff of fumes an idea wafts over me. Sitting there on my bike and

---

[9] This footnote has nothing to do with the title of this section, but I wanted to jump in here in the early going and mention something. In fact—and not to get all meta on you —this footnote is actually about the footnotes you're going to find going forward. See, I've read enough of this type of book to know what a dense chore they can be, and the big idea with this project was to take that entire concept and flip it on its head. I want to tug on your coattails about something, yes, but it's my intention to do it in such a way that you don't mind so much, and you even have an OK time along the way. The simple truth is that I appreciate the shit out of you for picking up this book in the first place, and the last thing I want to do is bore you into a coma. So yes, there are going to be a lot of references down here, and even a few *ibids*… but there's also going to be some (hopefully) fun, interesting, and informative stuff down here, too.

[10] No, I don't ride the Hog thing: I'm a Duc chick. 848 EVO, murdered out. For those of you who follow that, cool. And if you don't and I have to break that all down for you: it's not important to the story, it's just me popping off a wheelie for no real reason. The point is I'm a "grab and go" kind of gal, not a human coatrack with a purse hanging off the crook of her arm.

tripping on some high-octane *gasolina* I'm rocked by a thunderous "a-ha" moment. The prospect of having to remove the ignition key, get up off my ass, strip off my gloves and gear, and insert the key into my seat compartment to get out my wallet *all* so I can swipe my credit card at the pump to get my gas was blowin' my road high and just generally annoying the shit out of me. Why couldn't I just sit, fat-assed, on my bike, keep my gloves 'n gear on, effortlessly retrieve my credit card from my phone case (my phone was in my pocket, not locked away), slide it in at the ol' Texaco, do the zip-code-digit-dance over the keypad, and be done?

It was late fall, 2007. The smartphone wars were just ramping up—the BlackBerry Pearl, Bold, and Storm were all the rage, iPhones were just cresting the horizon, and the world was on 3G fire—but users were already stowing credit cards behind phone skins, peeling back the case and tucking them in. Still, this was an imperfect solution: it wasn't exactly convenient, and it certainly limited the life of the case; I, for one, was already plenty sick of replacing skins that had split and/or ripped at the corners from my repeatedly peeling them back to get at my credit cards, and the bulky duct tape fixes I came up with didn't comply with the margin for error on my skinny jeans. Why, I thought, couldn't the phone skin have a credit card slot *built in*?

Remember what I said about multiple degrees? Yeah, that. I hadn't cured cancer, but I had stumbled on the first principle in a good invention: it meets a need you didn't quite realize you had.

Still, just because something is a good idea doesn't mean that the road rises to meet it. A *lot* of good ideas never make it past the initial whisper-it-to-those-you-trust-and-dream-about-how-great-it-would-be-if-it-was-real phase. There's a graveyard of great ideas out there that never had a chance: inventors who never got started, who ran out of steam and/or money, who got derailed by other things in their lives.[11] Me, I was lucky. My kids and my close pals—the buddies I rode with—were all big believers... or maybe these

---

[11] And that's not to mention the guy who was set to invent the next best widget but who died when his car was T-boned by a drunk driver, or the top team of AIDS researchers who were on their way to completing monumental advances in their research but who were all wiped out in the Malaysian Airlines crash in July of 2014...

latter were just as fat-assed and biker-lazy as I was. Either way, they were my first "Go for it, Kip" cheering squad. In fact, of everyone I eventually told, I never had anyone tell me that it was a "bad idea": the worst I ever got was the marginally patronizing, "That's a good little idea." I chose to focus on the operative words "good" and "idea": I honestly had zero ego about the thing and had no problem admitting that, in the grand scheme of things—and compared to the big ideas that were saving lives and/or improving the quality of life for the sick or aged or infirm—my idea was pretty "little."

Still, if it was a "good little idea" that could be universally adopted as the world went smartphone... Well, then, maybe we would just see who was calling what "little."

Granted, my explanations of the idea were better received than my first few sketches. I Jackson Pollacked a design on the back of a cocktail napkin a few margaritas in, trying to show folks what I meant by "a phone case meets a credit card wallet." I may as well have titled my happy-hour masterpiece "a smartphone walks into a bar." People looked at my drawing upside down and sideways, then smiled politely... Not really the response you're looking for on your elevator pitch dry runs. Oh, well. *Señorita? Por favor, un otro round...*

Truth was, though, I had bigger problems than artistic ineptitude. The combo "credit card wallet meets phone case" idea came to me right in the middle of what were some rather funky lean years: at the time I was 1) in the throes of a separation/divorce and 2) raising four kids solo in N.Y.C. Having an idea was all well and good, but pushing that idea from cocktail napkin to patent-pending prototype was going to take more money and more time and energy than I had to spare. It was already all I (and my close personal friend, Red Bull) could do to keep all the plates I had spinning from wobbling off their perches and crashing down on my head. Still, I wasn't ready to give up the ship just yet. Why? The only answer I can give is: it probably has something to do with whatever it is in me that makes you never wanna' cut in front of me in traffic. My tendency to get all "dog with a bone" on a thing wasn't letting up, despite the external pressure to just *puh-lease ruh-lease*. It was just obvious to me that this "good little idea" was too good to go unrealized, and that it would be someone else's if I didn't hold on. And pardon my Latin, but *irrumabo*[12] that. It may have been only a "good little idea" but it was *my* good little idea, God-damnit, and I wasn't letting go. I was going to hang onto this towrope even after I'd face-planted in the boat's wake and my bathing suit was seining the ocean.

My very first prototype came into being in June of 2008. If you're imagining a machine shop, and guys in hard hats adjusting equipment to spec... don't. The first iteration of my idea involved a billfold wallet, a phone skin, and a roll of duct tape.[13] I laid the billfold on, wound that tape around my phone skin, slid in my

---

[12] Latin for FUUUUUCKKKK.

[13] Quick inventor shoutout here: in my opinion duct tape is, hands down, the best invention ever, and its creator deserves a far-too-rarely-given round of applause. In the 1940s Vesta Stoudt, a mother with two sons serving in the Navy, went to work in the Green River Ordnance Plant between Dixon and Amboy, Illinois to do her part to help her sons and their fellow servicemen. Vesta got a job inspecting and packing the cartridges that launched the rifle grenades used by soldiers in the Army and Navy. The cartridges were packed 11 to a box, and the boxes were taped and waxed to make them waterproof. The box flaps were sealed with thin paper tape, and a tab of tape was left loose so that it could be pulled to release the waterproof wax coating and open the box. The problem was that the thin paper tape wasn't very strong, and the tabs frequently tore off when soldiers pulled on them, leaving these soldiers scrambling to claw the boxes open while under enemy fire. The paper tape put lives at risk—including the lives of Vesta's sons. So Vesta came up with a solution: seal the boxes with a strong, cloth-based waterproof tape instead of the thin paper tape. Vesta raised the issue with her supervisors but, though they thought it was a good idea, she didn't get anywhere close to having it implemented. So Vesta did what any mom with two sons in the Navy would do: she wrote a letter to President Franklin Delano Roosevelt outlining the issue and explaining her idea about how to fix it. Vesta said: "I suggested we use a strong cloth tape to close seams, and make tab of same. It worked fine, I showed it to different government inspectors they said it was all right, but I could never get them to change tape." (Letter from Vesta Stoudt to President Franklin Delano Roosevelt, February 10, 1943.)

Roosevelt approved of the idea; he sent it to the War Production Board, who wrote back to Stoudt: "The Ordnance Department has not only pressed this idea... but has now informed us that the change you have recommended has been approved with the comment that the idea is of exceptional merit." (War Production Board's Ordnance Department letter to Vesta Stoudt, March 26, 1943.)

The War Production Board tasked the Revolite Corporation with creating and producing the product, and Stoudt received the Chicago Tribune's War Worker Award for her idea and her persistence with it.

credit cards, cash, and ID, and *voilà*. Best part of all? My design was skinny-jeans-user friendly.

And maybe not wanting to get up off my bike to pump my gas makes me the laziest person on earth, but it turns out that I had enough get-up-and-go to get off the couch and go see a man about a thing. My tennis buddy Victor (and his crushing forehand return) worked for a plastics manufacturing mogul—I'll call him "Carl"—and he introduced us. I went into our meeting and I said the four words more likely to elicit an eye roll than interest in a man who heard them a hundred times a day: "I got an idea." However—lo and f'ing behold—Carl actually *liked* my idea! He liked it so well, in fact, that he got his mechanical engineers to draft up a CAD drawing[14] of an injection mold so that his Shanghai factory could run off a bunch of prototypes. He said, "I'm the kind of person who believes that a product in hand is better than an idea sketched out on a scrap of paper." As it turned out, he was also the kind of person—and the only big-deal businessman I've met to date—who really believes in doing a deal on a handshake and his word. We left that meeting with an understanding that, for a percentage of the company, Carl would put the whole weight of his substantial operation behind my idea. I'd just secured production.

As you can probably imagine, having someone like Carl in my corner gave me a serious boost. Here was a guy with the means and the know-how to make my cocktail-napkin idea a plastic-and-silicone reality. Hell, his company made containers and caps and tubes and bottles for most of the cosmetic industry. Whenever some *faaaaaaabulous* designer wanted to manufacture a cap for her eco-friendly, go-green perfume with some exotic potato-based polymer, Carl and his team were the ones who figured out how it could be done and delivered. I felt sure that the product he and I came up with was

---

[14] "CAD stands for Computer Aided Design (and/or drafting, depending on the industry) and is computer software used to create 2D and 3D models and designs." (https://www.smartdraw.com/cad/) In my case, the drawings designed the injection molds that would be used in production.

going to be absolute dynamite, and I felt stupendously lucky to have Carl on my team.[15]

The prototype-manufacturing process had a three-to-five-month turnaround and I didn't really know what to do in the interim, but it was suggested (read: strongly recommended) that I keep my mouth shut and my head down, and that I go find a patent lawyer to help me protect this idea *right quick*. It was good advice, and I guess it should have been obvious from the get-go: it doesn't take a genius to realize that, in a world on the brink of a smartphone revolution, in which practically every new smartphone will need a case, a combo phone-and-credit-card case could mean big money, and you'd be best served to *shuttupay-ouface* real fast unless you want someone else to eat your lunch.

So a-hunting I went for a patent lawyer. I'd list the whos, whats, and where-fores, but the whole process is a blur (within a blur, within a blur…) of subway rides, rushed meetings, and phone calls to and from secretaries' secretaries' secretaries… The point being, I truly cannot now recall exactly how I arrived at the patent firm I finally did: can't tell you exactly how I ended up on the top floor of that über-posh Park Avenue law firm situated at the corner of "everything you've got" and "you look faint." Too embarrassed to say that I must have gotten off at the wrong subway stop and that no, there was no way I could afford their legal services, I sat and watched as this patent lawyer, in his crisp Brooks Brothers suit, grew all animated about my "good little idea."

"Fantastic," he told me. "We'll draft and file the patent. It's crucial that we do this quickly, given the universality of the idea."

To which I thought, "OK, I guess I can swing funding this somehow, because as long as I have the strength and expertise of this law firm behind me then I'm surely in good hands."

---

[15] Can you sense the other shoe about to drop? Then you're already two steps ahead of me, and about a million miles ahead of where I was at that moment. The lessons I learned in that particular outing are included in Part V, in the rundown of all the many pitfalls and blind alleys I walked into and down on this "adventure." For now, stay tuned.

What do they say about assumptions? Let me tell you: "surely" is a dangerous word.

Mr. Brooks Brothers was all smiles as he handed me the papers to sign, and His Girl Friday was nice enough to offer me a (very expensive leather) seat at the (long, single-slab mahogany) conference table and a glass of water from a (Waterford Crystal) decanter when she saw the look on my face as I walked out of his office. I tried to smile at her, but I felt more like a stroke victim. What had I just signed away? My life's savings? My first born? I felt ill. Still: I had representation. At the very least, I felt (mostly) confident, I was doing what I was supposed to do.

Someone once described the patent process to me this way: Imagine that you're living in America in the early days of westward expansion. You're looking to have a place all your own, and you hear that out west of the Mississippi there are huge swathes of unsettled land available to anyone willing to claim a piece of it. In fact, someone has even organized a process by which such claims are made and registered. You show up at the appointed time and place, out on the edge of the continent's great central prairie, with hundreds of others just like you. You're handed a stake and told that, when the starting gun sounds, you're to run out into that great big expanse and "stake your claim" to the piece of land you want. An official will be by directly to note for the record that this parcel now belongs to *you* and no one else (and *certainly* not to the Cheyenne or the Apache or the Pawnee: after all, if they wanted it so bad then why didn't they register with the office?).

Uncomfortable glance back at our nation's not-so-spotless record on fair play with property rights (and *plenty* more on that later) or not, this little peek back through the sands of time gives us a pretty good metaphor for the simple idea behind the patent system. A patent filing is essentially equivalent to our predecessor's stake in the ground: it's a way to assert and register that "this here is my claim, and my stake to prove it."

The metaphor is apt in more ways than one. As I said before, I had no ego

13

invested in my idea, and I certainly didn't have enough *chutzpah* to think that I was the only person on earth to have an idea as simple as a combination wallet and phone case. As with a lot of "good little ideas," the race was going to come down to who got up off their ass first: who drafted the designs, filled out the forms, paid the fees, and filed the sucker before the rest of the hopefuls had a chance to get off the mark. Like with our land-grabbing westward pioneers, it was going to be all about who got there first.[16]

Here's the thing about filing a patent, though. It ain't like the movies. Your patent application will be viewed—maybe, if you're lucky—within *three years*. If that's just too long to wait then you can speed up the process—for a price—and slide into what the patent world calls the "expedited patent track." I guess we have different definitions of what the word "expedited" means, the patent world and I: once we opted for this option, the only thing that seemed to go any faster was the money out of my little startup company's bank account.[17] Still, the expedited process skipped us ahead a year in the patent processing line. Not nothing, I guess, but it still felt like trying to run in that nightmare where you're stuck... in... slow... motion...

---

[16] Note to any patent scholars and/or litigators reading this: Yeah, I know it's more complicated than this, but we've got to start somewhere, don't we? And what is a long discussion of first-to-file vs. first-to-invent, and which standard was used when I filed my patents, and when it changed, and how the margins become blurry between first-to-invent and first-to-file anyway, because proving you invented something before the guy who filed first can be tough and open to interpretation, making it something of a *de facto* first-to-file system anyway, going to get us? It would just confuse the nice people with a lot of granular legislative and procedural details they don't need, at this point. We'll get there, I promise. For now, I need you to bear with me.

[17] Oh yeah, and did I mention? Filing a patent is *expeeeeeeeeeensive*. Totals vary based on different factors, but for a good general breakdown you can check out the article "The Cost of Obtaining a Patent in the US" by Gene Quinn of IP Watchdog. According to Gene, for an "extremely simple" invention you're looking at $5,000 in attorney and filing fees at a minimum; for a software-related patent you're looking at $16,000+. So bear in mind: just because your idea is "little," it doesn't mean the cost will be. Check out the article at https://www.ipwatchdog.com/2015/04/04/the-cost-of-obtaining-a-patent-in-the-us/id=56485/

And the thing is, Young Grasshopper, that for every second of those two years you get to worry and fret over whether someone else has filed before you: get to worry and fret about whether you told a friend who told a friend who told a friend who's a patent attorney and who, with a swift swoosh of his power tie and his mighty pen, drafted an application for your "good little idea" and left you right off it; get to worry and fret about whether your patent and all the work those lawyers did over long liquid lunches at your expense will come to anything at all. Absolutely nothing is guaranteed, and all you're left with is questions to which no one has satisfactory answers. Did we get there first? Was our application submitted properly? Was it drafted well, not too vague but not too specific (because yes, cross my heart, either one can knock you right out of contention in the already-crowded IP sphere)? All you can do is keep swimming, little fish, and hope that the sharks don't smell your fear (though if they do your lawyers will certainly step in to help... at their usual rate of $800 per hour, of course).

I did get one thing done, at least, in the midst of all that waiting: I finally came up with a name for my contraption. And who knows? Maybe it had something to do with the fact that, at that time, my ongoing dealings with lawyers (both patent- and divorce-related) were putting me constantly in mind of ocean-dwelling predators, the kind who rend and maim before they kill, who seem to sense weakness and feel no mercy... or maybe not. All I know is that, rising suddenly from the warm depths of REM sleep, I scribbled "CardShark" on the back of a random scrap of paper I had on my nightstand. I woke up in the morning bleary-eyed and wondering where that name even came from, but thankful: the name was perfect, and it meant that one more piece of the puzzle had fallen into place.

Other pieces were soon to follow: our prototype arrived, got a quick mod' (Carl grabbed the scissors off his desk and tweaked the cutaway thumb tab), and headed back to the Shanghai plant with a substantial order attached. In fact, by November of 2008, I felt like I was well on my way: I'd filed everything I'd been told to file and had "patent pending" status, I had a great name and now, with the prototype turned around, I had inventory on the way. I think I even allowed myself a celebratory lil' nose crinkling smile.

"I did it!" I thought. "I'm off to the races. I'm good to go."

Not so fast, little fish. Not so much. Not. At. All.

II. THE ANSWER TO 99
OUT OF 100 QUESTIONS

Let me cut in for a minute here so we can have a quick chat about something that'll make just about anybody sit up and take notice. To quote Tom Cruise's character (who is, in fact, quoting Tom Cruise's character's father) in Cameron Crowe's 2001 film *Vanilla Sky*: "What's the answer to 99 out of 100 questions? *Money*."

What's the big deal, here? We're just talking about a little phone case idea, right? Well, yes and no. Since 2007 smartphone use worldwide has absolutely exploded, and this surge in usage and sales has had a direct positive impact on the demand for all the accessories that go along with smartphone ownership. What do I mean when I say "exploded?" 1.4 million iPhones sold in 2007 ballooned to 11.6 million in 2008, while BlackBerry producer RIM saw its sales increase from 6.4 million to 13.8 million in that same timeframe. It didn't stop there, either: in 2009 RIM had its best sales year ever with 26 million phones sold, though Apple was close on their heels with 20.7 million iPhones sold. The tide finally turned in 2012 when Apple surpassed BlackBerry, selling over 125 million iPhones to BlackBerry's 49 million.[18] Altogether in 2020, smartphone

---

[18] "BlackBerry phones could be gone for good as last major firm stops making them," William Gallagher, Apple Insider, February 3, 2020. https://appleinsider.com/articles/20/02/03/blackberry-phones-could-be-gone-for-good-as-last-major-firm-stops-making-them

vendors sold around 1.38 billion smartphones worldwide,[19] and that number is only expected to increase: smartphone use in the Indian market alone is expected to rise 84% by 2022, from 468 million (2017) to 859 million users.[20] That's a lot of phones… and a lot of phone cases… and a lot of money to be made by someone in a position to meet the growing demand.

By the time I got into the game, the major players in the space were already operating and well established: Case Mate, Speck, Incipio, OtterBox, and others were already selling their solutions to the inevitable moment when (not if) you dropped your phone, scrambled its guts, and shattered its screen (leaving you all *cacas de fortunas*[21]). Still, nobody was doing what I was doing: in fact, as far as I could see, by offering both protection and wallet functionality in a sleek (read: skinny-jeans friendly) design, the CardShark was set to unseat the biggies and claim the throne in the smartphone case market. I even—I'll admit it—felt like I *deserved* it. I'd had a great idea and I'd done everything right. I'd sunk everything I had into it, had taken a big risk. I'd partnered with people who could make what needed to happen happen. I'd even—at the suggestion of my are-you-f'ing-kidding-me-with-this-rate law firm—filed additional patents to further cover variations on the design.[22] I'd "expanded my patent portfolio"

---

[19] "Smartphone sales worldwide 2007-2021," S. O'Dea, Statista, March 31, 2021. https://www.statista.com/statistics/263437/global-smartphone-sales-to-end-users-since-2007/

[20] "Smartphone users expected to rise 84% to 859m by 2022: Assocham-PwC study," ET Bureau, Economic Times, May 10, 2019. https://economictimes.indiatimes.com/tech/hardware/smartphone-users-expected-to-rise-84-to-859m-by-2022-assocham-pwc-study/articleshow/69260487.cms

[21] Latin, yes, for "shit outta luck."

[22] For those of you wondering, the applications I filed were all for utility patents. A utility patent "protects the functional aspects of an article, i.e., the way the article works and is used, whereas a design patent only protects the ornamental appearance of an article, such as its shape, configuration and/or its surface ornamentation." (*Red Points*, "Utility Patent vs Design Patent") As it was explained to me, a design patent is fairly useless in terms of IP defense unless it is issued to a company that can afford to substantively counterattack the more powerful utility patent, should it find itself at odds with one. Meaning that, if you're looking to file yourself, the utility patent is the priority. As my first patent lawyer told me, "The utility patent is the big lily pad in the pond. A design patent is a smaller lily pad and sits on your bigger utility patent lily pad." *Ribbit ribbit.*

thinking that, once these patents were approved, at worst I could sell the whole thing off to one of the big phone case manufacturers and put my kids through college on the proceeds. And, despite my posh lawyer's warning that "there are no guarantees against someone else's patent filing predating yours, and knocking your patent out based on an earlier filing date"[23] (this despite the fact that he'd just cashed my "all your worldly possessions" check and accepted my first born as collateral) I had every reason to believe that I'd gotten there first. I'd staked my claim and was set to corner the market. I'd positioned myself perfectly—or, at least, as perfectly as anyone could—and now I was going to reap the reward for all the time, effort, stress, and expense, and the many, many sleepless nights.

And then...

The first combination credit card wallet/phone case that I can remember seeing was in the Verizon Store. Saying that my heart sank would be putting it mildly: I felt downright sucker punched. I pulled the case off the shelf, inspected the box and the package, read the description... It felt like a nightmare and déjà vu at the same time. This was *my* description of *my* credit card wallet phone case, but *this wasn't my product.*

I called my lawyer. "What does this mean?" I practically shouted into my phone. "Did they beat us to the punch? Did they get the patent before us?"

To which my über-calm[24] lawyer assured me that no, they did not, and yes, I was still the one primed to hold the patent in this space. *Whew...* sort of. His words were cold comfort in the face of the sleekly-packaged phone case I was holding in my hands. A little better was his assurance that, once my patent was

---

[23] Remember the land grab analogy? All filings are sealed until they're considered for approval two to three years later, and the fact that no similar patent existed when we were doing our due diligence before filing was no guarantee that someone else's application wasn't sitting ahead of ours in the queue like a manilla-enveloped grenade, waiting for the examiner to pull the pin and explode all of my indie inventor dreams...

[24] Of course he was calm: it wasn't *his* invention staring back at him. And anyway, it was all billable hours to him... but more on that later.

approved, we could (theoretically) seek recompense for the infringement retroactive back to the date of filing—a fact I came to lean on pretty heavily over the months that followed, as more and more of these combo cases started arriving on the scene.

And… guess what? It turns out that I *did* get there first. Somehow, despite everything, despite the other players in the space, I—Kip Azzoni Doyle—managed to secure the first patent on the combination credit card wallet/smartphone case. My lawyer summoned me up to his office, told me to sit down, and Vanna White'd me the patent application, number 8,047,364,[25] with that beautiful and officious "Validation" stamp on the top. I could finally peel the duct tape off my mouth, take a deep breath… and scribble out another big fat check to my only-too-happy-to-help lawyer. The amount I signed over damn near gave me a seizure,[26] but even still: I'd done it. I had my patent. I'd joined the ranks of the bona fide indies.

And I feel like I should note, here: People in the industry were *shocked* that I managed to secure these patents. Eric Hurwitz, my current representative (and certified IP minefield ballerina[27]), told me that when he first read the summary on me and my patent he was sure he'd read a typo. It was, he said, a very big deal. (And thank goodness: if it wasn't, then Eric might never have come onboard, and then where would I be?)

So now's the time when we have *the talk*. No, not that talk: if your parents

---

[25] In a flurry of exuberance I actually thought about running out and having that number tramp-stamp tattooed… Given everything that followed, I'd say I dodged the laser-removal bullet on that one. Tally that as another lonely hashmark in the sparsely-populated win column…

[26] You don't know what you don't know, and at the time I didn't know that I could have saved myself some serious money if I'd filed with a small, Main-Street-U.S.A. patent law firm, rather than the top-of-the-world law firm I used. Ma'bad, but at least now I can go out and make bank on my invention… right? Right?

[27] Don't worry if that doesn't make sense: you'll meet Eric later, and all will be revealed.

or siblings or the older kids in middle school or internet porn didn't give you enough information to confuse you more than you already were then you're *cacas de fortunas*; this book may be about the little guy getting screwed by the big machine, but I don't think you're going to find it particularly instructive for your purposes. No, the talk we need to have is about *monetization*. You had a good idea and you got it protected, but how do you turn that good idea into cold, hard cash?

Really, you'd be wise to define your business strategy right up front. It's that Roger Daltrey moment: "Now tell me: Who the fuck are you?" You'd be wise to ask yourself questions like, "Am I going to manufacture this product myself? If so, am I going to line up a manufacturer in the U.S.? Or am I going to find a manufacturer in China or Mexico, because here in the U.S. it takes five workers to do what one 12-year-old Chinese girl working overtime into the next millennium can do?"

It's an honest question, but its rhetorical aspect is something that the maker of any new product needs to consider. If the bottom line is *the* determining factor (and how can it not be? What new startup is so capital-flush that it can make operational decisions purely on principle?) and no meaningful impediment stands in the way, we *will* ship manufacturing overseas in absolute deference to that lower per-unit cost. It might not be high-grade globalization theory I'm dishing here, but it's worth the comment: consumers only see the shiny end product; they don't have to think about how the sausage gets made, and that includes what happens to factory-line fingers and limbs in a country where dinner often consists of "anything that has its back to God..." but you do you.

And while you're chewing on that, consider this: if you don't, someone else will. At some point it was made clear to me that, while I might choose to manufacture the CardShark here in the U.S.A., the (inevitable) counterfeiters certainly wouldn't. Their low production costs would allow them to significantly undercut me on price, and I'd end up priced out of the market on *my own damn product.*[28] Me and my principles would be out of business and out on our asses in no time flat.

So let's say you run that all through central processing and you make the call to manufacture overseas. The 12-year-old Chinese girl may lose a digit or two, but your mother taught you not to stare. Thank the nice Chinese girl for the low, low production costs and move on. It's like when you're 20-something and drunk and you don't tip the towel gal in the nightclub ladies' room at 2 A.M.... Will it ever really matter? It's not like you're gonna see her again. This Chinese girl, she's not in your backyard. You don't have to actually *see* her finger stumps. You don't have to look her in the eye. And hey, maybe you can still assuage any lingering guilt by having your product "Assembled in the U.S.A.," which is a nice way of saying that some 10-fingered American dropped your product in a box in South Carolina before shipping it off to parts beyond... Ah, the dirty business of global market capitalism. The thing is, of course, that if you live in the developed world you had blood on your hands before you knew the name of the game. You had blood on your hands before you even knew

---

[28] Words of wisdom from my super astute husband, Mark Doyle, who was/is a forensic accountant for the U.S. government, and who was embedded in Afghanistan for a year with the anti-corruption task force, living and working in a container to track dark money: he calls this phenomenon "racing to the bottom." His words: "While seemingly a good idea for the consumer, the effect in the marketplace ends up being devastating for U.S. manufacturers and their employees. The logic is that if you have low or no minimum wage constrictions it helps keep payroll wages under control, and you can employ so many more dirt cheap. The broader problem is that if you have an economy that is constructed on the basic premise that paying workers the least amount you can pay them will lead to a 'stronger economy,' then you have to ask yourself, 'Stronger for whom?'" I'd also like to take this opportunity to point out that Mark's business *Rags of Honor*, a silkscreen and apparel company, uses "Made in America" t-shirts, hires homeless and at-risk veterans, and pays them a higher-than-minimum wage. Their operating principle: "They had our backs, let's keep the shirt on theirs." That's how Mark pays it forward and thanks these men and women for their courage and service. This is my big shout out and unapologetic plug for Mark's tireless work and his fantastic team, who I am proud to work alongside and call family. Mark has never taken a dime out of this company: he built it so that folks who want to help but don't know how can actually do something. Buy a shirt and help a veteran at www.ragsofhonor.us.

that you were playing, and of all of the privileges associated with living in the First World the most damning of all might just be the luxury of the ignorance of this fact. Decide to buy this particular ticket and take this particular ride—decide to bring a product to market—and you'll be freed from this particular illusion. And what then? You can shrug and write it off as "just the way things are," but you'll no longer be able to pretend that you don't know.

So you buy the ticket and you take the ride. You make your peace and you dirty your hands. You'll square your conscience later; for the time being you've got more pressing concerns. Namely, now that you've got your product in hand: how are you going to compete with the big-ass players in the space (and their legion of lawyers), the ones who have the infrastructure and the business relationships to get their (infringing) version of your (legitimate) product placed front and center on big-box-store shelves?[29] After all: you may be the rightful holder of the patent on this product, but no one elected Walmart arbiter of that dispute. If you've got an issue with the manufacturer you need to take that up with them: otherwise the name of the game is *sales*. If your competitor, licensed or not, infringing or not, can deliver more product with better sales and a better markup, then you, little indie, might be *cacas de fortunas* all over again. And good luck suing that infringing competitor when all of your working capital is sunk into inventory you can't move...

This is probably the right place to point out that internet sales and Amazon fulfillment are another verse in this same song: for you it's all about cost and visibility, and if you're up against an infringer who can afford to *go yard* on its SEO and targeted marketing then you may find yourself relegated to the "bottom shelf" of the internet just as much as you would be exiled from central placement in the big-box stores. Click-throughs fall off drastically the farther down you are on the list of Google search returns, and God help you if your product lands on the second page. Time to drag some rocks down to the beach

---

[29] And realize: if your product is any good, this *will* happen.

and spell out "S.O.S.," because, Wilson, you need *hhhheeeeeeeeeeeellllllllllpppp*.

And look at that: it's another in a series of opportunities to take a long, hard look at yourself and decide, hoo boy, just what you want to do. Do you want to micromanage every single step of this project, all the way from invention through manufacturing to warehousing and sales and distribution, only to fight the big companies who will inevitably gobble up the prime shelf space at the top retail chains? Do you *really* want to do all that? Isn't there another way? Well sure there is, little indie! Read on!

Occupying their own strange middle ground in the producer-to-consumer universe, QVC, "As Seen on TV," and the Home Shopping Network (HSN) offer indie inventors the chance to get their nails and hair did, work out so they look real good, dress it on up in rhinestones and pumped-up heels, and work their best game-show-hostess hand sweep as they pimp their wares on camera. It's its own universe and it comes complete with its own Bible (and yes, they do call it that) of rules, regulations, conditions, and sins one must never commit on QVC, "As seen on TV," or the HSN.[30]

It's a complicated process, getting your particular widget on one of these outlets. With QVC, your first meeting is with the gatekeepers of the outer circle: these are the guys who came up with such answers to modern living as the ShamWow and the Snuggie, the veritable living legends of the genre, who've parlayed their QVC successes into a company that acts as a go-between for aspiring indies and the QVC powers-that-be down in some Godforsaken place in Pennsylvania. Once you get past these guys (for, I should add, a percentage of any and all sales, should your product make it to the big show) then you get handed off to the next tier... and the next... and the next... each of which takes a bite.

---

[30] This Bible also explains 1) everything you're expected to do—out of pocket—to bring your product to market on their platform, and 2) that all of your pallets have wheels on them, meaning that whatever doesn't sell comes back to you... for a full refund, of course.

All of which is to say that, with these routes, you're playing the thin margin game, and it's a very specific needle you have to thread *vis a vis* your price minus your costs plus these platforms' standard cut(s) to make these avenues work for you at all: otherwise you're essentially paying to play. Which can work, if you think that the exposure is going to lead to more sales down the road that will offset these losses; for me and the CardShark, though, the numbers just didn't work. If it does for you, though, I wish you luck. Feel free to come on over and borrow anything that fits. I'd be glad to help get you ready for your closeup.

So where does that leave you? Or, more specifically, where did it leave me? I had a certain amount of inventory from my partnership with Carl, but it was also becoming clear to me that 1) moving units myself was a slow, expensive, thin-margined, and fraught way to do things, and 2) infringers were almost guaranteed to infringe, and at a unit volume that was beyond anything I had the infrastructure to match. Thank you so much for giving me yet another opportunity to know myself a little better, to stare into the mirror and lip-sync those sweet, raspy words: "Aw who the fuck are you?" Specifically, am I a non-practicing entity (NPE), someone who makes money from the licensing of her patented IP but who *doesn't actually make anything herself*? Is that what I set out to do? Is that who I set out to be?

The question is more complicated than it initially appears, for reasons that will (I hope) become painfully clear to you as we slide through some pertinent history in the next part of this book. For the moment, however, let me just say that the patent system itself, in its current iteration, is somewhat[31] antagonistically inclined toward NPEs and their dark shadow, the Patent Troll. In fact, recent legislation has given the infringers some serious teeth against the infringed in the form of what's called an "inter partes review," or IPR. And what's an inter partes review, you ask? It's just a means for a large, infringing company

---

[31] And this year's winner of the "Understatement of the Year" goes to...

to challenge (and potentially invalidate) the patent holder's patent(s). It's just a means for a large, infringing company to cast the patent holder into can't-pursue-infringement-litigation purgatory for the duration of the 18-month (at a minimum) proceedings. It's just a way for a large, infringing, and deep-pocketed company to bleed the little guy dry. It's just one of the (many) ways that the little fish are being devoured by the big corporate carnivores in today's IP ocean.

...but much more about this in the sections to come. Suffice it to say, for the moment at least, that even in this sit-back-and-get-paid approach, in the modern IP sphere, safety is not guaranteed.

So what's left? You could always make the patents themselves the product: build your portfolio, sell it and get out, and buy yourself a charming little getaway on a remote island in the Quirimbas Archipelago. Watch the waves roll in, sip away on a daiquiri, and wonder if you have any more great ideas in you. Maybe write a book about the whole adventure and rack up some royalties while you catch some rays. More than anything, *enjoy yourself*. Enjoy being you. You're one of life's big winners. Isn't it nice? Or how about, instead of doing that, you *WAKE UP* from that daydream and find that you've run out of runway: that you're leveraged to the hilt and every avenue to profit and solvency has been systematically usurped or whittled down to a pinhole; that the debts incurred in processing, filing, production, promotion, litigation, *et al* have buried you up to your neck and the tide is coming in; that you just can't keep it going anymore. Not much left to do now but sell your IP to the first "patent-monetization" company[32] that comes along with a lowball offer. This company doesn't intend to actually manufacture anything with your patent, so there's no point in negotiating for royalties from future sales: no, they're going to use your patents to squeeze money from infringers. They're going to go fight what would have been your battles, if you had the capital to float the expense.

---

[32] Read: "Patent Troll."

If they get a licensing deal for their trouble, more for the good. If they get a settlement, who's complaining? If they get IPR'd and (once your, now) their patent gets invalidated… *eh.* Cost of doing business. They bought the thing for a song anyway, and it's just one of the dozens (if not hundreds, if not thousands) they'll run through this same playbook… Trust me, they'll come out all right in the end. And anyway, no matter what happens with them, it'll be all the same to you, babe. That ain't your IP no more. You sweated, bled, spent, and tore yourself to shreds to bring your good little idea to patent-protected life… and now it's all over.

As I would come to learn, there are many ways to die out here in the IP jungle; this, however, may be the ugliest.

Of course, sitting in my attorney's office with the elation of "Validation" in my heart and the sting of his last bill still cramping my check-writing hand, I didn't know any of this. I'd had my toe in the water for a minute, now, had been around the block a bit, but—compared to where this whole thing would go in the coming years, and what I would learn when it went there—I was still babe-in-the-woods brand new. Looking back, I can honestly say that I didn't have a clue what I was in for… not that I'm sure it would have mattered if I did. Remember what I said about you never wanting to cut in front of me in traffic? Indies are a quirky lot: if we weren't scrappier than your average bear we wouldn't make it past the initial obstacles to become indies in the first place. Now fully patent protected, and with even more money I didn't have to spend sunk in and gone, I was more invested than ever.

Either way, whatever I knew or didn't know, and whether I intended to become a full-fledged NPE or not, step one of Operation: Make Kip Some of Her Damn Money Back was to *lock that IP shit down*. Starting from the moment of approval, not to mention retroactive to the point of filing, the other combo cases on the market were infringing with a capital "I," and our first move was to send out some friendly letters letting those infringing know that they were flying in my patent's airspace: that they needed to either cease and desist or take a meeting to discuss how much they were going to pay to license (read: keep making money off of) my IP.

Step two? Get the biggies onboard. Apple and BlackBerry's parent company RIM were the biggest players in the field, at this point, and were already in the phone accessory game: a license from either (or both) of them could mean big things for CardShark, LLC. I drafted and sent along a pitch and a rough description of the CardShark, along with a (standard) non-disclosure agreement (NDA) for them to sign if they wanted to see the particulars and details. And—thrilling for me and my dreams of college tuition payments and early retirement (Quirimbas Archipelago here I come!)—both companies seemed excited to see the CardShark. I even got a nice note back from the head of Apple's accessories division who raved, "Great invention, I've CardSharked my iPhone!"

"Golly gee willikers," I thought, already smelling the ink on the royalty checks. "The guys at Apple are using my invention! They've even verbed it! They must really like it! They're going to take a license for sure!"

Again: not so fast, little fish. I didn't know it at the time, but I was about to enter a whole new phase of WTF… and earn myself another expensive lesson in lawyering in the process.

As I said, my lawyers had worked up an NDA for me to send along with my pitch letters. "Here's a bit of thigh," my pitch letter said, "but if you want me to open my kimono, then you're going to have to sign the enclosed." A signed NDA was another guarantee that the company couldn't run off and build a product on their own based on the materials I'd shown them. It meant that my idea was doubly protected… or at least that's what that mob of by-the-hour lawyers told me.

So with patents in place and NDAs on hand to fire off at a moment's notice, I'm making calls to all the big-name phone case manufacturers. I've got my pitch down pat, ready to wow the first human I get on the line. I've got a call in to one particular company, one of the top phone accessory manufacturers. I'm navigating their phone system like a pro, shaving time on the corners. "Press '1' for directory, '2' for sales, '3' for..." I don't even wait. I know what I'm after. I arrive at some lowly employee, start my broad pitch. Everything's going smooth until I get to the part about the non-disclosure agreement—the part where they need to sign on the line before I dump the secret sauce in their laps.

"Nah," the employee tells me. "Sorry, we don't operate that way."

Huh? What? You... don't... operate that way? Are... you... KIDDING ME? "It's an NDA, and everyone has to sign it!"

Jesus, did I say that out loud?

I guess I did, because the lowly employee at the big phone accessory company responds with the phone-tone equivalent of an eye roll to reiterate that said big phone accessory company does not sign NDAs, that it would be impossible for them to do such a thing, and that, Oh by the way, while we're on the subject: they are a massive phone accessory company employing hundreds of people who live and breathe in the phone accessory space and have for years, and it's staggeringly likely that they, the big phone accessory company, have already thought up my idea in-house and have already decided whether (or not) to pursue it.

Oh, really? Want to tell me who died and made you Alexander Graham Bell?

Bottom line, the employee told me: "You can submit your idea to us, but we reserve the right to produce it or to decide whether it fits our company's core design ideas."

Not much of an offer, is it? But when you're a little fish in the great big ocean of the booming phone accessory market, maybe you're not in a position to negotiate. And anyway: a patent's a patent, right? It wasn't like they could just go off and produce the design without taking a license, right? (Right, guys? Right?) So in what I can only describe now as a hallucinatory bout of trust, good faith, and flat-out optimism, I opened my kimono. I drafted the email and I bared all. I felt like I had to; I felt like if I didn't then there would be no chance of getting this industry biggie to consider licensing my product idea or (hope hope hope, fingers crossed, mama needs a new pair of shoes) full-on buy out my patent portfolio. I made sure to let them know that I was also sending along (attached) a letter from my lawyer stating that "if you ever go into this credit-card-wallet-smartphone-case world, we retain the right to come knock on your door..."[33] And I crossed all my fingers and toes and I waited.

And I waited. And I waited. And then I waited some more. Finally, a form-letter-style response from the phone accessory giant hit my inbox. They'd carefully considered my CardShark walletskin designs, they said, but had finally decided that they were "not going in this direction" with their cases. They thanked me for giving them the opportunity to consider it, and wished me a lot of luck in my future endeavors.

Other shoe, prepare to drop.

At some point subsequent to my bout of ill-advised trust and optimism I was tipped off by a very good source that this industry biggie actually had a pretty sketchy reputation for ripping off ideas. Still, when I actually got the call, it knocked the wind clear out of me. The "call" in this case, was a photo texted to me by my 14-year-old daughter: she was walking to school, downtown N.Y.C., when she happened to look up.

"Hey, Ma," the accompanying text read, "have you seen this billboard?"

Staring back at me from my phone screen was my CardShark idea, larger than life, brought to you, the smartphone users of the world, by none other than my good friends at the big phone accessory company. You know: the ones who'd crossed their hearts and hoped to die that, sorry, they just weren't interested at all.

Up until this point I'd been in steady contact with RIM and Apple. Now, though, and all of a sudden, I found that I was being ghosted by both of them.

---

[33] Fat lot of good it did me. Later, this email exchange would be the only proof I had that I'd shown this company my invention. The fact that I needed said proof should tell you exactly what's coming next.

Still, I told myself not to worry: they were just busy, I thought. They'd get back to me soon. Everything was still fine. And then the word came down: my friends at the big phone accessory company had won their bid, and were partnering with both Apple *and* RIM. The biggies were going into partnership with my invention, and I was nowhere near it. The big accessory company had swooped in and stolen my sale like a pickpocket in Grand Central.

They say it's the hit you don't see coming that levels you. While this shot didn't put me out, I don't mind telling you that I took the full standing eight count. Making things worse was the fact that I'd *just* moved ahead with a 5,000-unit order for the current iPhone model based on our positive response from Apple. We figured that any product that got "verbed" by Apple's accessory department execs had to be a sure winner, and we wanted to have the product on hand when the inevitable order came. Now, instead, I was stuck sitting on this inventory and watching from the sidelines as the big accessory company's knockoff populated into Apple stores and Apple carrier websites: stuck watching from the sidelines as their bootlegged version of my product became an instant top seller.

In the interval between the billboard drop and the eventual sit-down at the arbiter's office oh-so-many months later, I probably could (and maybe should) have been cited for harassment. I was on the phone to my lawyers constantly in a desperate attempt to understand just *what the actual* was going on, and just *what the actual* they were doing about all the aggressive activity around my IP.[34] I told them I needed answers, told them I needed action, told them that if they didn't get something done *pronto* I would "cease and desist" with their services... to little appreciable effect. To be fair, maybe, I'm not sure that anything short of Federal Marshalls storming the big accessory company's offices, raiding retailers shelves for this contraband, would have satisfied me. I wanted *immediate impact*, shock and awe, and by the close of business and that just

---

[34] For the record: my lawyer's counsel that "imitation is the sincerest form of flattery" was not, as you can maybe imagine, appreciated.

ain't what lawyers do. They assured me that they were working on it, told me I had to be patient, promised that what needed to happen would happen… while all the while a little machine on their end counted up the seconds of the call, tallying up the billable hours. Who knew that words could cost so much? Thousands of dollars later, and—despite their promises and their mandate to aggressively defend my invention—the only thing I had to show for all this time spent was the bill.

That… and a growing list of infringing products and outlets where these rip-offs were being sold. Because it didn't stop with our friends at the big case company: once the concept hit the market and proved a hit every other manufacturer, legit or not, jumped on board to cash in… as did every major retailer. Everywhere I went—Bloomingdale's, Nordstrom's, Best Buy, Staples, Verizon, Apple, T-Mobile, Sprint—I found combo cases staring back at me from the shelves. I saw them hanging on pegs at mall kiosks, saw them in those phone-accessory stands in airports. I was harassed down in Chinatown by street vendors peddling the cheapest CardShark ripoffs around. My superb team of research assistants—a.k.a. my scatter of four kids, each with their own social circle—were on the case, too: together they uncovered more foul play than all of the lawyers combined. My oldest daughter, Kalypso, by then in college, almost got thrown out of class for snapping a pic (during her psych exam) of her classmate's *fer shit* wallet case.[35] Each time I or one of my kids spotted an infringer I called my lawyer to let him know… and each time the little machine clicked on to add the seconds to my ever-increasing tab.

If you haven't already picked up on the fact, you can ask any one of my former (or current) lawyers and they'll be glad to tell you: nobody moves at the pace I want them to. Which is to say that my mounting frustrations with the apparent lack of action on the part of my legal team didn't mean that nothing was happening. Machinations were underway, a whole lotta' lawyerin' was go-

---

[35] My kids: pure, unadulterated awesome, and the kind of next-level reconnaissance I could have never ever paid for.

ing on, and it all finally came to a head when we were summoned to an arbitration (read: "settle with" versus "sue") meeting with big infringer's learned counsel.

It's Christmastime, now, so I subway it uptown and skirt the rich tourists teaming under the iconic snowflake at 57th and 5th Avenue to yet another overblown law firm on the corner of 58th and *faaaaabulous*. I elevator on up to the 51st floor, inspect my faux-corporate look (not too "I am woman, hear me roar," nor too "Hello boys, come hither") in the polished-steel doors, and step out into the lobby where my lawyer is already waiting. I'm still wondering if I can pull off the Rambo-meets-a-rainbow vibe I'm going for when the secretary tells us that they're ready for us, and we can go in.

Now, a quick question for you: If it's a school of fish, a pod of whales, and a murder of crows, then what do you call a group of lawyers? My vote's for "slaughter," because that's what it felt like we were walking into. Situated around the large conference table are no less than six lawyers hailing from various locales and coasts, all there to put the kibosh on our infringement claim. And facing them... Kip and her lone lawyer. The heavies at the big infringer have brought an assault rifle to a thumb war.

Still, despite the home-field advantage and the six-to-one representation imbalance, the arbitration is not the quick and brutal affair that I, despite my best efforts to remain optimistic, fear it will be. In fact it drags on and on: 10 hours in we're still there, hashing. (At least we get a lunch break: the turkey clubs the secretary brings in look delicious until it occurs to me, sandwich in hand, that the catering for this affair is going to be paid for by the loser of the proceedings.) Finally, though, the slaughter cuts to the chase and "drops the big one": they have prior art pre-dating my patented design, they tell us, and will take matters up with the Patent Trial and Review Board[36] and get our patent invalidated outright if we don't *step off.*

---

[36] The dreaded PTAB, where IPRs are adjudicated, a.k.a. where patents go to die. We're going to talk about this a loooooottttttt more in the coming pages. For now, make a mental note that it was from the lips of these corporate bullies that you heard it first. That should set the appropriate tone for understanding all that follows.

IP vocab 101: Prior art is "any evidence that your invention is already known. Prior art does not need to exist physically or be commercially available. It is enough that someone, somewhere, sometime previously has described or shown or made something that contains a use of technology that is very similar to your invention."[37] Basically big infringer's lawyers are telling me that they have artwork showing that the big infringer got there first, that they drafted the idea before me, and that I don't have any right to get mad[38] about them using their own idea... an idea that, sorry, was never really mine in the first place.

Except... how about... no. Bullying ain't gonna budge this biaaatch. I've been through too much just to get us here. If you think I'm going to roll over now, you've got another thing coming.

The head of the slaughter is still smirking at me across the conference table. Now he sighs his most patronizing sigh, like he's exhausted by the sheer weight of all the needless trouble I've put him through, and the effort of being patient and understanding through it all.

"Clearly, Miss Azzoni," he says, "you don't know how things work in the patent world."

And then he smiles a full, wide-mouthed smile. Like he's my friend. Like he just wants us to dispense with the unpleasantness and get back to how things were before I caused all this ruckus. And sure, I look perplexed... but it's not for the reason he thinks. It's not about him and his slaughter questioning the validity of my (legit, valid, signed, sealed, and delivered) patent; it's not even about his choice to get braces when he's already so old. No, I'm perplexed by his timing. Because if he's got prior art proving that my patent is invalid, then *why the fuck have we been sitting here for 10 hours?*

"Huh," I say, all valley girl uptalk. "Gosh, I guess... I must not? Since...

---

[37] "What is prior art?" European Patent Office. https://www.epo.org/learning/materials/inventors-handbook/novelty/prior-art.html

[38] My mom used to say "dogs get mad, people get angry." Let me tell you: I was mad.

like... I'm the one with the patents? And you're the ones ripping me off? And... um... why are we here after 10 hours if you have proof that my patents are... like... invalid? Why did you wait to drop this bomb until 5 P.M., instead of doing it first thing this morning?"

That's right, even Barbie got her PhD.[39] Their "big one" is a fucking dud.

Still, by the end of the day, their intimidation tactics are wearing me down. When the slaughter head changes tack and leans in to tell us, straight-faced and shameless, that his multi-million-dollar-company client is "cash-strapped" and "can't pay me all at once for all back damages due," I want nothing more than to buckle into one big shoulder-shaking sob. Here I've done everything right: I've kept my mouth shut, covered my ass with a sheaf of patents, and hired a top law firm to help me navigate and defend myself in the world of product invention and innovation. I've been flagrantly ripped off by one infringer after another, and now this infringer—with its hand in the cookie jar—wants the CardShark, a flailing startup in dire need of the money owed for these back damages, to become their *fucking lending institution*? Not today, sunshine. I'm pushing back from the table and gearing up to tell the lot of them where they can shove their poverty claims but my lawyer, I guess sensing that I'm gonna' geyser, executes a strategic "pull aside."

"Look," he tells me, "we've got them. Why don't you go outside and get some air, and let me handle the details?"

Sure. Fine. Great idea. I walk back out into lobby, and he stays to cut the deal behind my (supremely pissed-off) back. At least, I think, we've won. Whatever haggling is happening behind those big conference room doors pales in importance, in that brief moment, to the broader point: we've notched a tally in the win column. I'd had a significant brush with some seriously unscrupulous shit, but now, it seems, the tide is finally—and rightfully—turning.

---

[39] "Computer Engineer Barbie Has a PhD In FUN (And Breaking Down Stereotypes)," Brian Barrett, Gizmodo, February 12, 2010. https://gizmodo.com/computer-engineer-barbie-has-a-phd-in-fun-and-breaking-5470587

And then... *flash bang*. My lawyer comes out, all smiles, to let me know that he's settled with the slaughter... *for a pittance*. For a tiny percentage of the revenues the company has accrued during the time they've been ripping off my invention. A meager amount that they will pay out over an extended period of time—a meager amount that will, in fact, go straight into my lawyer's account to pay down my back balance. He doesn't even let me clear it through my CardShark bank account for the sake of my taxes first. *I never even see it.*

Turns out my lawyer had a good reason for that strategic pull aside: he didn't want the deal to blow up. He wanted *his money*. This is the same lawyer who told me, "Kip, you keep waiting for the other shoe to drop because when it does it's going to be on their feet, not yours. Just relax." Lawyers.[40] Their words are so useless, and yet each one costs so fucking much.

And that was it. The slaughter of lawyers took private Towncars back to tri-state airports to fly to their scatter of cross-country and coastal cities, while I scraped together coins and picked shards of stale gum from the bottom of my bag for subway fare back home.

Sigh. You ever have one of those moments when you really wish you could just Frankenstein yourself in a brain from someone smarter? One of those moments where you can see that you've paddled right up to the edge of your intelligence, and that the parts beyond are reserved for Ivy Leaguers and Mensa types? I may not have all the smarts in the world, but most of the time I have enough smarts to know when I'm out over my skis: enough smarts to know what I don't know and that, when I don't know, I should defer to those who do. In this case, I knew that the patent system was there to protect the work of indie inventors like me: the rest, I figured, I could leave up to my expensive expert lawyers. Just goes to show you that sometimes, even when you think you know that you don't know, you still don't know.

---

[40] I have nothing to say about lawyers. Nothing nice, that is. Comedian Steven Wright said it best: "99% of lawyers give the rest a bad name."

As it turns out, my lawyer only doing enough lawyerin' to cover his own interest in the matter was only *part* of the behind-the-scenes two-step going on that day. Ladies and gentlemen, meet Denise Pierce.[41] Denise was part of the slaughter, one of the six who pushed back from the conference table that day. Given that 1) I had other things on my mind, and 2) I figured (read: hoped) I would never have to see any of these "people" ever again, I never thought twice about her smart Chanel suit or her matching I-just-killed-a-whole-cow for-this-size purse. But whereas the other five flew off to existences utterly (thank God) disconnected from mine, Denise didn't. Oh, no: Denise wasn't done just yet. She showed up again, in the next dispute with the next infringer the Card-Shark went after. And then she showed up again with the next. Turns out Denise was scouring the docket for CardShark filings, and then offering herself to these infringing companies as an expert in the space: someone with an intimate knowledge of the CardShark's patents and—more importantly—our arbitration strategy and settlement history.

And let me tell you: whatever these companies were paying her, she was worth it. She squeezed and we buckled.

Why did we buckle? Because we were in a corner, and Denise knew it. Because the unfortunate truth about justice—the unfortunate truth I had to learn *right quick*—is that it is *fucking expensive.* If you have the resources to fight for it then you can have it... maybe... but that's a *very* big "if." When it came to Denise, she knew just how far into deep water we could swim, and at what point we would start to drown. If they wanted, these infringers had the resources to drag any proceeding out and keep us tied up in courts, hemorrhaging filing and attorney fees. And sure, they had these same expenses... but what's a few hundred thousand to a company making millions? Especially when that few hundred thousand saves millions in damages that will never have to be paid if things go Denise's way.

---

41 Names changed to protect the slimy.

A pattern started to form: the infringers would infringe, we would object, they—with Denise at the helm—would threaten to drag us through seven levels of legal and bureaucratic hell, we would settle, and Denise would move on... right into the waiting arms of the next infringer.

"Shit," I started to think, "for all the new Chanel suits and matching bags that CardShark dealings have bought her, maybe I can at least get some advertising out of this whole thing. Maybe throw a big 'CardShark' logo across the shoulders of one of those Chanel suit coats like on a NASCAR jacket. See if I can make this situation work for me."

Alas, no such luck. Unbranded, Denise continued to weave her way through the never-ending game of infringer Whac-A-Mole that had become my life.

Bad as I thought all of that was, it paled in comparison to what followed. Next stop: The Rocket Docket of East Texas.

That's right, Texas: home of the chainsaw massacre.

Like before, this story starts with an infringer. This company—let's call them Company X—was ripping me off, and we had proof that I'd flashed open my kimono and shown them all my secrets. Out goes our cease-and-desist letter: *Dear infringer, You're a thief. Quite stealing my shit. This is my idea, not yours. Either cease and desist at once or license my IP and pay me a royalty. Until that time, "get offa' my cloud." Kisses, Kip.*

Remember how I said that "a whole lotta' lawyerin'" went on between our initial cease-and-desist letter to the big infringer and our final sit-down with the slaughter? Same story with each of the infringers that followed, and—while that might sound complicated and technical, and while it certainly cost me an arm and a leg—it mostly looked like this: We would write a letter informing the company that they were infringing. They would respond by either 1) not responding, or 2) full-on blowing us off. We would then send another letter which would, in all likelihood, produce the same result. We would continue our letter-writing campaign until eventually the infringer would respond by

37

100% denying that they were infringing: respond by saying that they were, in fact, incensed and outraged that we would even suggest such a thing. We would then kindly offer to discuss an amicable resolution to the situation, i.e. how to convert their infringement into a license. If they didn't want to do this, we would tell them, we would be forced to bring a suit against them. Then arbitration ("Hi, Denise, love the new outfit..."), then settlement (pittance), then on to the next one.

Or at least, that's how it usually went. In the case of Company X, though, things took a slightly different turn. This company received our cease-and-desist letter and cut right to the chase: What if, they wanted to know, instead of licensing our IP, they simply bought out our entire patent portfolio? We'd walk away well compensated ($5 million was floated as an estimated sale price), they'd build out their product line under rightfully-owned IP, and—as it would be their problem, now—we'd never have to track down and haggle with another infringer ever again.

We were headed into the Christmas season again, and for one brief, magical moment it felt like the universe was placing a big ol' present marked "To Kip, with Love" under my tree. Oh, the things I could do with $5 million. Of course we're interested, we told them. We'd be happy to chat. And while we're on the subject, sorry we came at you so hard. We didn't know that you were such a nice guy, just trying to do the right thing...

Intoxicating niceties followed. It was starting to feel like we might even hold hands, might get a few drinks and just see where the night takes us, and then... ghosted. Weeks went by with the suggestion on the table but no offer on the hook—weeks in which this infringer kept steadily making bank on stocking-stuffer sales of their unlicensed walletskins.

Quite the spot for the little CardShark to find herself in! What should she do? Should she push for action, knowing that too much eagerness—or aggression—might sour the deal? Or should she wait and see, knowing that this all might be a ploy to kick licensing and litigation down the road while her rightful earnings flowed into Company X's coffers? What to do, what to do... and here it is, the holidays. The season for charity and goodwill, for giving the benefit of the doubt, for seeing the good in your fellow man.

"They're probably just busy," the little CardShark thinks to herself, ignoring the still-stinging memory of what happened the last time she got ghosted. "And after all, it's a big decision. $5 million is a lot of money. They probably just don't want to rush into anything."

Yeah... no. Not so much. Extension after extension runs out, until my lawyers finally tell Company X the jig is up. In the immortal words of George W.: "Fool me once, shame on you. Fool me twice... we won't get fooled again." Take a license or we'll see you in court.

All hand-in-the-cookie-jar, Company X agrees. Send the papers, they say, and we'll get it done. Out goes the agreement for them to sign: the terms, the license, the settlement for back damages and royalty payments going forward. It ain't a cool $5 mil, but it ain't nothing: and at least, with those back damages forthcoming, the holidays sales we waited through won't be a total loss.

With the matter all but settled, my thoughts finally turn to the season at hand. Ah, Christmas in New York. The snow falling on the City, ice skating in Central Park, the rich tourists with their Goddamned bags. Oh, well. At least— with Company X settled and the back damages and royalties on their way—it looks like it's going to be a very happy new year.

So just imagine my surprise, then, when instead of a signed licensing agreement we receive from Company X an 11th-hour about-face and FUCK YOU lawsuit served at 5 P.M. on Christmas Eve. And where, you might not think to ask, is this lawsuit being adjudicated? Why, the Lone Star State, of course: Company X's home turf. And don't you worry neither if none of this makes any Goddamned sense—if an infringer suing *you* is about the most up-side-down thing you've ever heard of: You'll have plenty of time to think about that on the plane ride down. For now it's time to get your spurs on and giddy-up, Cowgirl: your ass is sued in Texas.

What. The. Actual. FUUUUUUUUUUUCCCCCKKKKK.

No one to call, now. No more lawyers at their desks. They're all long gone, the contrails misting in the wake of their first-class flights to beautiful, expensive Christmas vacations. And here I am with a lawsuit posting deadlines early in the new year. I'm going to need to find and hire a Texas-practicing lawyer to counter this fiasco before the end of day... today. Today of all days. Christmas. Mother. Fucking. Eve.

Ho, ho, ho.

Confused? It took me a lap around the mulberry bush to get it straight myself. Turns out Company X's lawyers are smarter than they dressed... or at least they knew the house rules. In particular, they knew that in the Eastern District of "Yeehaw" Texas—within which our plaintiff, Company X, was situated—the CardShark's cease-and-desist letter constituted sufficient grounds for Company X to turn around and level what amounted to an allegation of *extortion* against us. The story went like this: one day they were just going about their business, working hard to make an honest dollar and feed their broods of lil' Texans, when all of a sudden we dropped in out of the clear blue sky (and from New York City, no less!) with a threat and a demand for money. We'd brought the threat to their hometown, and they were prepared to stand before a judge and ask him to make us cease and desist any and all threatening activity in this a' here jurisdiction *pronto*.

Never mind that we were the wronged party, and never mind that we held the patents and the rights associated with same.

How can such a lawsuit even happen? Which is to say: How is it that I, the holder of a fully validated and protected patent, in this day and age and in this great land of ours, can't write a simple cease-and-desist letter to an infringer in Texas to inform them that *they're ripping me off* without opening myself up to a lawsuit in the process? Call me old fashioned, but I'd say that anything you can't defend ain't exactly *protected*.

It turns out that, at least in principle, the case law agrees. I'm going to turn to Marketa Trimble and her 2010 paper "Setting Foot on Enemy Ground: Cease-and-Desist Letters, DMCA Notifications and Personal Jurisdiction in Declaratory Judgment Actions," for the details:

> *When [patent holders'] rights are infringed in a foreign forum, rights holders typically get a chance to "set foot on enemy ground" without being attacked there. Without subjecting themselves to personal jurisdiction in the infringer's forum, rights holders can send cease-and-desist letters and inform the infringer of the right holders' rights, identity, and intentions to enforce the rights. Rights holders are protected from personal jurisdiction in the infringer's forum because courts, recognizing the importance of the letters in the policing of rights by rights holders and the seeking of settlements of disputes, do not consider such letters to be sufficient bases for personal jurisdiction over rights holders. Therefore, if the alleged infringer files a declaratory judgment suit as a counterattack to the letters* [Kip's note: This is what Company X did to me]*, courts in the alleged infringer's forum will not extend personal jurisdiction over a non-resident rights holder based only on the fact that the rights holder sent a letter to the alleged infringer; the sending of a letter by itself will not be a sufficient basis for personal jurisdiction. Promoting the use of letters in this manner is believed to advance judicial economy, and although this approach to cease-and-desist letters is not without its critics, case law is well settled on the point.* [42]

In the Eastern District of Texas, though, what is "well settled" is still open for interpretation, and the burden for filing a declaratory judgment suit is a

---

[42] "Setting Foot on Enemy Ground: Cease-and-Desist Letters, DMCA Notifications and Personal Jurisdiction in Declaratory Judgment Actions," Marketa Trimble, Scholarly Commons at UNLV Boyd Law, 2010. https://scholars.law.unlv.edu/cgi/viewcontent.cgi?article=1114&context=facpub

little easier to meet. Company X was trying to make the case that our sending a cease-and-desist letter *did* subject us to jurisdiction in their forum under the prior mandate: a precedent set by a Second-Circuit decision in 1997[43] held that when a rights holder sent cease-and-desist letters to an infringer he had technically "conducted business"—and thus made himself subject to jurisdiction—in the infringer's forum. Additionally, as Trimble explains in a footnote to the above: "[I]f the [cease and desist] letter is accompanied by other acts committed by the rights holder in the forum jurisdiction, or if the rights holder engages in regular activities there, specific or general jurisdiction may be extended over the non-resident rights holder based on such acts either alone or in combination with the letter or activities." The fact that we'd discussed and referenced royalties, damages, and the potential sale of our patent portfolio, in the generous estimation of the Eastern District of "The Stars at Night Are Big and Bright," meant that we'd potentially gone beyond the pale when it came to the definition of "conducting business."

Oh, and by the way: this particular lawsuit? She was a *beaut*. The opposing counsel made a case that read more like a lawsuit against me personally, going so far as to assert that I'd changed the corporate ownership structure of the company in order to do business under some obscure "alter ego." I wish I was kidding. What I'd actually done was create a parent company for the CardShark so that, if something happened to me, the CardShark would be owned by the parent company that my kids would then inherit. The CardShark, at least, would live on, and my kids would (hopefully) enjoy the fruits of my labors. However, according to the lawsuit filings, the parent company was my attempt to fake a shell company. Kip, as described in the lawsuit, sat back in the shadows, pulling the strings of corporate schemes; she careened around the globe without a care in the world, jetting off to fabulous and exotic places on the back of her fat company expense account. (To which I wanted to scream: "Oh,

---

[43] PDK Labs, Inc. v. Friedlander, 103 F.3d 1105, 1109

yeah? You think if I were jet-setting around the world I would have jet-setted my ass down to the Godforsaken Eastern District of "Remember the Alamo"? Do ya? Do ya really?")

The lesson here is how to spend another $15,000 you don't have to defend yourself against a frivolous lawsuit when all you want to do is defend your trampled patents against those doing the trampling. The suit may not have come to anything in the end—it was inevitably thrown out for improper juris-diction—but it certainly added significantly to the list of hassles, headaches, obstacles, and expenses that stood between me and a settlement... which was the whole point. Company X admitted after it was all over that they had sued me because they knew that, when they didn't sign our settlement agreement, I was going to sue them, hauling their asses to my jurisdiction—N.Y.C.—in the process, and they didn't want to pay for any of it. They knew that their lawsuit would fail: they'd just figured that, if they sued me first, they might bleed me enough to make me go away.

It's like that thing about a tree falling in the forest: If you have a boxing match and you lose a decision, but the other guy dies in the dressing room after the fight, who really won? We weren't down and out just yet, but it felt like we didn't have too many of these near-misses left in us. It felt like we'd tiptoed through a minefield down in Texas, and the Big Dream—of pursuing the in-fringers, converting them to licensees, and with the earnings from the quarterly royalties pivoting to manufacturing, promoting, and selling the CardShark ("the Original and Still the Best Smartphone Walletskin in the Game!") myself—felt farther away than ever. I was keeping faith alive, but barely. And on top of that, with each passing month and year, my patents' lifespan was running out.[44] Something needed to happen soon. I needed things to turn my way.

---

[44] Oh yeah, did I forget to mention? A patent ain't forever. A U.S. utility patent has a term of 20 years from its earliest effective date. And also did I mention? Maintenance fees are due at three and a half, seven and a half, and 11.5 years after issuance or the patent will expire at four, eight, or 12 years respectively. Put that in your been-down-so-long-it-looks-like-up-to-me budget and smoke it.

And then...

A new infringer's ads started popping up on my socials media feeds. Silly me I clicked on one, thus inciting the persistent wrath of the targeted marketing Gods. In this case, the suddenly ever-present ads were a near-constant reminder of how successful this company was becoming *off of my idea*. They were killing it, selling out of inventory over and over again. Finally I'd had enough: I pulled the trigger and fired off our letter. Ask not for whom the bell tolls, infringer: it tolls for thee! Cease and desist *right now*!

The owner, it turned out, was another *femme entrepreneur (#*metoo you guys!), L.A.-based and willing to meet. I was already heading out to L.A. for some other projects I was working on, so we set it up. Strangely enough, the meeting was to take place in the conference room of a law firm I'd once engaged for some copyright work. Oh well: no use crying over spilled lattes, eh? And anyway, this was just a friendly meeting between us gals, right? Sure. I should have listened to my buddy John, who once told me, "Remember, out in L.A., 'Trust me' means 'Fuck you.'" We'd gotten down to business, with pertinent papers and folders handed around, when all of a sudden the owner's husband barged in, grabbed the folder closest at hand, ripped it, and threatened the living shit out of me.

"I will destroy you and invalidate every patent you have if you come after her!" he spat. "We'll never pay you a fucking dime! I'd rather pay to IPR you!"

Startling? The kind of thing you'd like to just take a step back from while you get your bearings and try to figure out what, exactly, is going on, and how we went from "amicable sit-down" to "dramatic entrances and flat-out threats" in 0.01 seconds? Definitely. I'm with you. So why don't we go ahead and hit "pause" on this story—why don't we leave him standing there with his eyes all slitted and his teeth all bared and the spittle still on his lips—and talk about something else for a minute. Something unrelated, something pleasant. What do you say we talk about one of my favorite pastimes? After all, inventing and fighting infringers isn't all I do. Everybody needs a way to unwind after a stressful week. Some people golf, some people crochet. For me, nothing clears the mind like a few laps around the track. No, I'm not talking about running: I'm talking about pulling on a helmet and strapping into a four-point harness for a few high-speed turns. The racetrack might be one of my favorite places on earth. Out here, an entirely separate set of skills and awarenesses kick in. You don't hesitate, because out here hesitation kills: a game of inches becomes a game of millimeters becomes a game of skidding out into the sidewall when you're trying to fit your front-to-rear-bumper span inside the ever-shrinking space between the car ahead and the one who's rocketing up from behind while you all barrel toward the 3.0 hairpin turn that kicks into the "widow-maker" chicane. In this moment, most of all, you must engage: load the front tires with braking or load the rear tires with accelerating but neverneverever float between. Float between and you're neither coming nor going, neither here nor there. Float between and you're along for the ride, out of control. Like hitting black ice, but a thousand times worse. And maybe things slow down for you, then. Maybe the adrenaline perforating your central nervous system causes time to telescope in a new and interesting way so that you find a moment to wonder how you got here, how you went from "OK" to "impending disaster" so quickly, how things went so, so wrong. And maybe that's what happened that day in L.A.: maybe that was why I found the time—with him staring at me over the table and his threat still hanging in the air between us—to think back to that first meeting I had with my lawyer and his team after my patent validation finally came through. The question that day was what to do next, and the strategy the team advised was that we fire off our cease-and-desist letter to the seven most blatant infringers, what they called the "low hanging fruit."

"Why only seven?" I wanted to know. "Why not just send the same letter to all of the companies that are clearly infringing?"

And that was when my lawyer explained how the blowback from such a strategy could be potentially fatal: how, given that many of the companies that were infringing were large and deep-pocketed, and given that such an all-out frontal assault on all of them at once would give them a shared interest in the issue, this strategy would likely unify them against me and put them on the offensive.

"They'd IPR you," he told me. "They'd call the validity of your patents into question. They'd drag you back into the patent office and tie you up in proceedings and litigation until either the Patent Trial and Review Board invalidates your patents or you go broke. On top of that, they'll spread the cost around so that they never even feel it."

And why did I think back to that moment? Because this woman's husband wasn't just anybody. He was the president of something, something *big*. And no, I can't say what it was... but I also can't stop you from drawing whatever conclusions you want to draw from our proximity to the land of movie magic. And it occurred to me, then, why we were sitting in the conference room of this law firm in the first place. When you're on an insane retainer, you let the client—or the client's wife—do pretty much whatever he or she wants to do with your conference room, now don't you? And I could have pointed out that, as a former client of this firm myself, any such action would represent a conflict of interest... but I had the sense right then that there wasn't going to be much point.

I stood up and I left the meeting. I left the papers I'd brought on the table, and along with them I left everything this infringing company owed me in damages and royalties—north of $800,000, by my estimation, based on a deep-dive assessment of their sales-to-date. With his threat still ringing in my ears I swallowed hard and walked away from the biggest deal I would never make. I had to. I had no doubt that he meant exactly what he'd said: that legit claim or not, if I came after him he would IPR me and the CardShark until we were left with nothing at all... and that little expense would just be another item on his tax return. Who knows? He might even have found a way to write it off.

Oh, and this walletskin company? I still get their ads on my feed. It looks like they're still doing really, really well.

Ever have one of those dreams where you realize that you're dreaming? Where suddenly all of the strange and seemingly irreconcilable aspects that have troubled you for what feels like a long, long time seamlessly resolve themselves into the perfect realization that none of this is real—that this is just how things are in a dream—and the only thing to do now is to stop making sense and simply relearn the world? Walking out of that meeting into the bright, Los Angeles afternoon I had a sudden, clear sense of the world I thought I knew and understood falling away, and another one taking its place. In this new world, the fact that I'd done everything right didn't matter. The fact that I'd stayed the course, done the legwork, and footed the bill in the face of stiff odds and my fair share of personal hurdles didn't count for anything. In the upside-down Wonderland I'd stumbled into, the patent system—this thing I'd understood to be the very word-become-flesh of America's reverence for its inventors, its modern pioneers—was little more than an elaborate, cruel, and costly joke.

Which is to say: I was finally starting to see how it had been all along—or at least how it had been all along for me and the thousands[45] of indies who jumped into the IP game at the turn of the new millennium's first decade. Unbeknownst to me and, I'm sure, many other indies who were now waking up to the same harsh realities that I was, we'd jumped into the world of protected IP at a moment when things in that world were taking a turn for the worse for the independent inventor: a moment when corporate interference and the aggressive work of lobbyists were fundamentally rewriting the rules for how IP is handled and defended in this country: a time when—as my lawyer had tried to tell me all along—*safety is not guaranteed*. A time when, armed with the

---

[45] Check out https://www.uspto.gov/web/offices/ac/ido/oeip/taf/inv_utl.htm for a state-by-state and year-by-year headcount of Independent Inventors.

weaponized machinations of the new USPTO, an angry little man with more money than God could grow even richer from my IP... and there wasn't a damn thing I could do about it.

What's going on? I thought. What happened? Where did we go wrong? How did the system that's supposed to defend the little guy, the indie inventor—the system that's supposed to embody and protect the very heart and soul of the American dream, the notion that anybody with a good idea and a solid work ethic can make something for themselves—get so utterly and thoroughly fucked? And what does it mean for the future when well-heeled bullies—corporate or individual—can so thoroughly game the system for their own self-serving, ethics-flaunting ends? And what happens to America when all of the indies finally read the writing on the wall, shrug their shoulders, and give up trying because *there's just no way to win*? Who will invent tomorrow's duct tape and light bulbs and intermittent windshield wipers? The indies are and always have been a vibrant and vital and foundational element in the wild chemical reaction we call American Progress and Innovation: what happens to them *and to us* when the system that is supposed to defend them instead abandons them or—worse—becomes a weapon against them? And what can we do to save it, them, and ourselves?

In the coming years I would find answers to these questions, and in the coming pages I'd like to take you through them. Fair warning, though: the picture they paint is a bleak one. Bleak, I should say, but not hopeless, and by the end of this strange odyssey you'll have learned not only what can (and must) be done, but also the important part that you—yes *you*—have to play in it. For now, though, secure helmet and read on... it's going to be a bumpy ride.

PART II

CURIOUSER AND CURIOUSER

"I claim to be a simple individual liable to err like any other fellow mortal. I own, however, that I have humility enough to confess my errors and to retrace my steps."
—Mahatma Gandhi

"If you don't know where you've come from, you don't know where you're going."
—Maya Angelou

I'm doing my best here, but I'm certainly not going to say it any better than those two. The fact is that sometimes you have to look back if you want to go forward, and that's exactly what we need to do if we're going to have any chance of getting our bearings in the warped landscape of the modern patent system. I promise I'll do this as quick as I can: just the facts, only what you need to know. We can line 'em up and knock 'em back like shots. You may wake up with a headache, but—like with your favorite bartender—if I do my job right, you won't mind at the time.

49

I got help with this chapter and need to shine the spotlight on M. David Hoyle, an inventor who has had his patents stripped from him. What'd David do to deserve that? He invented something that the Big Tech companies wanted, so they rolled him. Let's just say that 95% of Google's platform sits on IP once protected by David's patents. Those patents are all gone, invalidated now, while Google enjoys and continues to expand the empire it built on David's back.[46]

The sad fact is, though, that in the bizarro world we now occupy, David's story is far from uncommon. David has tried to help me understand and explain how this could happen not only to him, but to so many independent inventors.

I've got to hand it to David: he still gets out of bed every morning and goes to work advocating for independent inventors against Big Tech infringers, despite Google's overwhelming success—to date—at distorting the truth. Frankly, I'm in awe of him. If I went through what he went through, I think you'd have to peel me off the floor. He's a true patent badass warrior, and we indie inventors are beyond lucky to have him fighting on our side.

So here we are at the cliff's edge. The water's deep, cold, and we're not sure what's swimming around under that inky surface. No turning back, though. We've come this far. Nothing left to do but plug our noses, count to three, and leap out into nothing. Deep breath, now. Prepare yourself for the *splash* and the sinking down. We won't be pushing up until we hit rock bottom...

I. THE SOUL OF AMERICA

"If a country can be said to possess a soul, then America's is the patent system:

---

46 Don't worry: I'll give David's story it's full due later on, in Parts III and IV.

50

*the simple, fair method of staking a claim to a new idea and getting the
chance to make money from it."*
—Julia Keller, *Mr. Gatling's Terrible Marvel* (Viking 2008)

We all learned in grade school that the United States came into being in
1776; what many of us didn't learn is that our split from the Mother Country
occurred right smack dab in the middle of the world's first major Industrial
Revolution. Beginning in the 1760s and building on what were largely British
innovations, this period saw a widespread transition across the modern world
from hand production methods to mechanized production, the increasing use
of steam and water power, the development of machine tools, and the creation
and broad implementation of the mechanized factory system. And it is no coin-
cidence that, at this time, England was the only country in the world to have
stabilized the regulation and enforcement of copyrights and patents: in England inventors could invent and, with protections in place, grow rich off of
their inventions. The incentive system was set to reward innovation, and the
industrious Brits were responding. By the time of the American Revolution,
England's influence was spread far across the globe: its industries had made it
the economic center of the world, and King George III ruled the seas with
ships built and furnished with patented British technologies.

As subjects of the Crown, the Colonists had access to these industries, technologies, and products; as an independent nation at war with the source, however, they were up the proverbial creek. No longer could Colonists ship cotton
back to England and have it returned in the form of cloth: now, without access
to England's industry, it took American textile interests six weeks to make one
set of sheets. It would have been an untenable situation for any new nation, let
alone a nation at war with the massively powerful British Empire. The new
Americans needed to step their game up fast, and a patent system—like the one
already proving itself across the pond—was a natural solution: a way to incen-

51

tivize American inventors to create new technologies for manufacturing, agriculture and commerce, and give legs to the fledgling country's aspirations for true self-determination.

...or at least that's how the story goes. Turns out there's a bit more to it than that. We talked a good game in our push for independence: the writings of this nation's founders lean heavily on the nobility and virtue of the undertaking. Still, we know now that the lessons we all got in grade school—the ones about George Washington's cherry tree and Benjamin Franklin's key and kite—skipped a point or two. Slavery, the displacement and genocide of Indigenous Peoples, endemic gender inequality: it's crystal clear that not everything was on the up-and-up behind those flowery words, and the world of intellectual property was no exception. In fact, while we were busy enshrining the sanctity of intellectual property in our own patent system, we were as busy (and knowingly) engaging in some pretty flagrant infringement of our own. The truth is that one very good but rather ugly reason our Founding Fathers felt so compelled to start protecting inventions in our new republic is because, like pesky teens nabbing a few Pall Malls out of mom's purse, we *juuuuuuuust miiiiiiiggggbhbht* have stolen a few key innovations on our way out the door and snuck them back Stateside with us. (Thanks, ma. Love you longtime.)

One very good—and historically significant—case in point:

In early September, 1789, 21-year-old Samuel Slater boarded a ship in London. He was bound for New York and the New World. He blended in, presenting himself up as a simple farm hand—which he was most definitely not. In his pocket he held official papers identifying him as a recently-released apprentice at a cotton mill.

And not just any cotton mill, either. Slater had apprenticed for seven years at the Cromford Cotton Mill in Derbyshire under Richard Arkwright, the patent holder of the water frame, a device which made it possible for thread to be spun on dozens of spindles at the same time and which was fundamentally transforming textile manufacturing. In fact, Slater had spent a great deal of time working with the water frame, as he had shown himself to be uniquely adept at maintaining and adjusting the machinery. In his seven years of apprenticeship he had plenty of time to study and understand each and every component of this patent-protected technology. On top of that, he had an exceptionally good memory.

Unfortunately for Arkwright, Slater wasn't just a good employee with a sharp mind. Slater was also ambitious. In 1789 he learned about the sorry state of textile manufacturing in the new America. Despite the fact that the new country was already one of the leading cotton suppliers in the world, the former Colonists was unable to do much more than supply bales of the raw material to processing centers abroad—processing centers which were growing rich selling the refined product to manufacturers and end users. Slater smelled an opportunity.

Despite the patent system in place, many technologies in England were still protected mainly through guilds and the tightly-controlled apprenticeships. Secrets about production were shared, but respected. By 1774, the English government was offering an extra set of protections to the textile industry— again, a main earner for the Empire—by criminalizing both the export of textile machinery and any implementation of textile mechanical knowledge outside of England. Meaning that, when Slater stepped off the boat in New York fresh from his stint at the Cromford Cotton Mill, with the intention to make his fortune on the back of his privileged knowledge, he was preparing to commit a criminal act.

Slater flashed his apprenticeship papers and was quickly hired on at a New York textile plant. He'd made the trip knowing that America's textile production was woefully behind the latest advancements, but it's hard to imagine that he wasn't a bit shocked when he walked into work and saw employees laboring over hand-operated machinery, using methods that must have seemed, to him, to be arcane, old-fashioned, and outdated. We have to wonder what advances he tried to suggest, and why they were not implemented; whatever they were and whatever the reason, his employment in New York was short lived. A few weeks later, upon learning about a manufacturer up in Providence who had been trying and failing to copy Arkwright's water frame, Slater sent off a letter explaining that he had significant experience operating and maintaining Arkwright's machine and offering his services. The manufacturer took a chance and brought Slater on as a partner, and Slater quickly went to work bringing the manufacturer's operations up to date with its English contemporaries. He worked from his excellent memory, making the necessary components himself, and experimenting and adjusting until he had completed his own version of Arkwright's original design. After a year of work, his machine was complete. America's first automated textile mill was operational.

Unsurprisingly the mill was a huge success, and the technology quickly spread. By 1815 there were over 130,000 spindles at work within the 140 textile mills that had sprung up within a 30-mile radius of Slater's mill. And from there it would continue: it's not an overstatement to say that Slater's mill marked the beginning of the modern American textile industry and the end of America's dependence on England's textile processing; nor is it hyperbole to say that it served as America's entry point into the First Industrial Revolution. Without Slater and his strategic IP theft, one wonders what would have become of this new and cash-strapped nation: wonders whether it would have managed to continue to stand at all, or if the ruling economics of the day would have eventually forced it back under the thumb of foreign influence and control.

Still, and despite the boon they represented to these new United States, Slater's case and others like it made it clear to our Founding Fathers that something had to be done to *lock that IP shit down*. After all: without clear protections in place, what was to prevent the next Samuel Slater from taking the fruits of *American* ingenuity back to England, or to some other as-yet-unconsidered emerging economy? In fact, two years before Slater's Atlantic crossing, on September 5, 1787, the Constitutional Convention had unanimously agreed to grant Congress the power to protect intellectual property: "to promote the

54

Progress of Science and useful Arts, by securing for limited Times to Authors and Inventors the exclusive Right to their respective Writings and Discoveries."[47] James Madison, known as the Father of the Constitution, was himself a strong advocate for patent rights: in The Federalist No. 43, published on January 23, 1788, he describes how Article 1, Section 8, Clause 8 of the Constitution is dedicated to the enumerated powers of Congress, and that expressly among these is the ability to *unify* the protection of IP rights, *secure those rights* for the individual rather than the state, and *incentivize* innovation and creative aspirations.[48] On January 8, 1790, during the first State of the Union speech to Congress, given only months after the ratification of the Constitution, President Washington himself implored Congress to pass a Patent Act to get the country moving with "new and useful inventions" and "the exertions of skill and genius in producing them." The Patent Act of 1790—only the third act of legislation passed by the young Congress—was the result of these efforts. Under this Act, the power to grant or refuse patents was given to three people: the Secretary of State, the Secretary of War, and the Attorney General. Applicants needed the consent of two of the three officials for their patent to be granted. An examination process would be carried out by the same three officials to determine whether the submitted invention was "not before known or used" and "sufficiently useful and important."

Just three years later, though, this 1790 Act was repealed and was replaced by the Patent Act of 1793. This Act was designed to rectify issues that had arisen from the process set forth in the 1790 Act. Under the prior Act, and given that the people responsible for examining and granting patents had other important duties associated with their appointments to attend to, the examination process often took several months: too long, critics felt. Likewise the defini-

---

[47] U.S. Constitution, Art. I, §8, cl. 8.

[48] "The Federalist Papers : No. 43," James Madison, January 23, 1788. https://avalon.law.yale.edu/18th_century/fed43.asp

tion—"sufficiently useful"—was problematic: who was to say what an invention's future use might be, and whether that use might prove sufficient? Accordingly, this new Act defined as worthy of consideration for patent protection "any new and useful art, machine, manufacture or composition of matter and any new and useful improvement on any art, machine, manufacture or composition of matter."[49] Inventions no longer needed to be "sufficiently useful and important": it was enough now that they were "not before known or used." The Act also significantly simplified the patent process: applicants needed only to petition the Secretary of State for protection of their invention; the duty of examination was assigned to the Attorney General.

In the three years between the 1790 Act and the 1793 Act only 57 patents had been granted; the 1793 Act, with its broader definitions, opened the floodgates on applications. Despite the new delegation of duties the patent office was still simply inadequate to consider and examine every one of the hundreds of applications pouring in, and accordingly, in the rush, patents were granted to inventions and processes which were neither original nor useful. More than this, whether through oversight or indifference, many "inventions" like Slater's made it through. During this era, in fact, as Pat Choate says in his 2007 work *Hot Property: The Stealing of Ideas in an Age of Globalization,* such "patents of importation" made America "the world's premier legal sanctuary for industrial pirates. Any American could bring a foreign innovation to the United States and commercialize the idea, all with total legal immunity."[50]

Another federal Patent Act was passed in 1836; as before, this new Act was designed to reform and rectify the issues plaguing its predecessor. Importantly, this Act created an official Patent Office, which up until that point had existed only as a duty within the Office of the Secretary of State, and created the role of

---

[49] This definition is the one still used by the patent office.

[50] "Hot Property: The Stealing of Ideas in an Age of Globalization," Pat Choate, Alfred A. Knopf, 2007. ISBN 9780375402128.

the Commissioner of Patents, whose duty it was to grant patents, freeing the Secretary of State from this duty. It also provided that newly-granted patents be made publicly accessible at libraries throughout the country, making it possible for potential applicants to check whether their invention was truly original before filing their application. This change greatly improved the quality of applications and, along with the creation of a dedicated office and commissioner, improved the quality and efficiency of the processing of applications. Importantly, the Act also made it possible for patent holders to file for extension of their protection: up until this point a patent holder's claim expired after 14 years; under the new Act, and for appropriate reasons, patent holders could appeal to have their protection extended for a further seven years. It also removed the U.S. nationality and residency requirements, making it possible for foreigners to file for U.S. patents.

The patent office itself wasn't completed until 20 years after the passage of the Act, though this did little to dampen America's enthusiasm for innovation or the protections afforded to them: once the office was completed, it quickly became the most visited office building in the country. Just two years after this completion, in his "Lecture on Discoveries and Inventions," delivered to various groups at various venues throughout 1858 and 1859, Abraham Lincoln himself reiterated to his audiences the vital importance of the patent process: "The patent laws began in England in 1624; and, in this country, with the adoption of our constitution. Before then any man might instantly use what another had invented; so that the inventor had no special advantage from his own invention. The patent system changed this; secured to the inventor, for a limited time, the exclusive use of his invention; and thereby added the fuel of interest to the fire of genius, in the discovery and production of new and useful things."

It should also be noted that Lincoln is still the only U.S. president to have been awarded a patent.

Still, and despite the advances made in the patent system, intellectual property protection and intellectual property infringement and theft always seemed to walk hand-in-hand. Along with the flood of applications, the semi-chaotic system that emerged from the 1793 Act created a seemingly endless supply of contested patents, and courts were overwhelmed with lawsuits over patent validity and infringement.[51] The Act of 1836, while allocating more resources to the matter, did nothing to alter the language and criteria for patent protection, and the tug-of-war between innovation and its protection and infringement from ambitious outsiders continued through the end of that century and into the next one, at once driving and threatening America's careening course through the Second Industrial Revolution.

II. THE GILDED AGE

This Second Industrial Revolution saw the advent of new technologies and inventions that dramatically changed the economies and the daily lives of people living and working in the United States, Europe, and Great Britain. Huge factories employing mechanization and automation technologies produced consumer goods at a rate previously unfathomable, new forms of transportation and communication effected a level of connection that was heretofore impossible, and mechanized farming turned agriculture into big business. The air brake, the typewriter, the electric light bulb, the tractor, the radio, as well as —importantly—the process for refining petroleum into its commercially-useable byproducts, laid the groundwork for America's emergence as an international center of industry and commerce in the 20th Century.

But within this flourishing, something dark and predatory was brewing.

---

[51] "The Patent Office: Its History, Activities and Organization," Gustavus Weber, The Johns Hopkins University Press, 1924.

Innovators were flourishing, growing rich from the fruits of their labors, and there were those who saw this and—like Samuel Slater before them—started dreaming of ways to profit from the innovations of others. Worse, certain of these had the position and the capital to skew this game of tug-of-war in their favor. These were the robber barons: businessmen willing to exploit, bully, ruin, and outright steal in the name of profit, expanded control, and their own general increase.

Again, a telling example emblematic of the times:

Nikola Tesla was a Serbian immigrant who arrived in New York City in 1884 with four cents in his pocket. What he lacked in means, however, he more than made up for in brains: he'd studied math and physics at the Technical University of Graz and philosophy at the University of Prague, and had spent two years repairing direct current (DC) power plants with the Continental Edison Company in Paris. Moreover, he was a brilliant inventor: he'd already come up with an idea for a brushless AC motor, a motor using alternating current (AC) to produce a rotating magnetic field, making the first sketches of its rotating electromagnets in the sand of the path where he was walking when the idea came to him. While AC electrification had been employed before, this idea represented a significant leap forward, as at this time there was no practical, working motor that ran on alternating current.

Beyond his ability, though, Tesla was motivated. He was obsessed with solving the riddle of the day: how to effectively and safely harness, distribute, and employ electricity.

While the early part of the 19th Century had played host to rapid and significant advances in the scientific understanding of electricity, the end of that century would see a similar race to develop methods for applying that understanding to the demands and aspirations of modern life. Up until this point, what industrial mechanization processes there were had been powered by shafts or belts that transferred the motion generated by hydraulic pressure, steam, or compressed air to machinery operating apart from those sources. Creating or modifying these exacting processes involved moving and recalibrating all of these various means of energy transfer, and was accordingly labor-intensive, time consuming, and expensive. The electric motor changed everything: with an electric motor each workstation or process could serve as its own independent source of motion, free from the belts and shafts that would have otherwise connected it to an external driver. This not only simplified the industrial process, it also increased productivity by allowing factories to equip their facilities in a way that took full advantage of the workspace.

At the time, the standard method for distributing electrical power was via direct current. Direct current runs continually in a single direction, from a power source to a power draw; because of the nature of this method, however, DC power is not easily converted to higher or lower voltages to meet varied demands. Tesla theorized that alternating current was the answer to this problem: alternating current reverses direction a certain number of times per second, and can be converted relatively easily to different voltages via a transformer should such variation be required by a mechanism or application.

Back then Thomas Edison was the biggest name in electricity, and his work, his patents, and the company he'd formed with the backing of robber baron JP Morgan—the Edison Company—all revolved around DC power. When Edison hired the young Tesla to work at his Manhattan headquarters, that work concerned DC power and the DC power generation systems that Edison favored. In fact, Edison is said to have offered Tesla $50,000 if he could improve on the design for Edison's DC dynamo generators. After months of work and experimentation Tesla presented his new design to Edison and asked for the money; according to Tesla, Edison refused to pay him, instead telling him, "Tesla, you don't understand our American humor." Unsurprisingly, Tesla quit soon after.

Tesla was undaunted. He found backers to support his research into alternating current, and throughout 1887 and 1888 he applied for and was granted more than 30 patents for his innovations and inventions. He was also invited to address the American Institute of Electrical Engineers, where he caught the attention of George Westinghouse, the inventor who had launched the first AC power system near Boston. Westinghouse was Edison's major competitor in what was already known then as the "War of the Currents," the competition surrounding the introduction and implementation of electric power transmission systems that was playing out between Edison's DC and Westinghouse's AC systems. Westinghouse, recognizing that Tesla's designs could be a powerful weapon in his efforts to unseat Edison's DC current, purchased Tesla's patents for $60,000 in stocks and cash and royalties on each horsepower of electricity sold. It would mark the beginning of a partnership that would run through some of the most important events in the history of America's modernization.

In 1893, the Chicago World's Fair was set to dazzle the world. This was a major cultural and social event: an elaborate presentation of American industrial optimism and the feats of modernization. With the flick of a switch, the entire "White City" was to be illuminated with electric light, and the job was up for bid to the "War of the Currents" combatants. JP Morgan, bidding on behalf of the General Electric Company—the company formed by the merging of the Edison Electric Company, in which Morgan was the main shareholder, and the Thomson-Houston Electric Company, in 1892—said GE could do the job for half a million dollars; Westinghouse, with a proposal employing Tesla's AC system, undercut that bid by half. Westinghouse won the contract and on the evening of May 1, 1893, President Grover Cleveland flicked a switch, illuminating the fairground's neoclassical buildings with the light from hundreds of thousand of bulbs and marveling the assembled crowd. It was the work of Tesla, Westinghouse, and 12 new 1,000-horsepower AC generators located in the Exposition's "Hall of Machinery."

The event marked a turning point in the "War of the Currents": it was clear to the 27,000,000 people who attended the Fair that AC was the power of the future, and from that point on more than 80% of the electrical devices ordered in the United States were those made to be run on alternating current. But with this success Westinghouse expanded too rapidly and too aggressively, and he left himself overextended and vulnerable. JP Morgan moved in to take advantage: the value of GE's DC patent technology was falling fast, and Morgan needed Tesla's IP—the AC patents now owned by Westinghouse—in order to keep solvent. In a move typical of the robber barons, he used his power, influence, and money to launch a smear campaign designed to topple Westinghouse's company. These efforts had their desired effect, and Wall Street responded with a run on Westinghouse stock. Westinghouse was on the ropes.

Westinghouse tried to interest investors, but they were wary. Tesla was certain that the patents Westinghouse now held on AC electricity would attract investment, but Westinghouse explained to Tesla: "The exact opposite is occurring. Nobody is willing to invest because of the royalty deal on your patents."

Fearing ruin, Westinghouse begged Tesla for relief from the deal. He told Tesla, "Your decision determines the fate of the Westinghouse Company." Tesla, grateful to the man for all he'd done for him and deciding that the benefits to society from his AC electrical system were more important than the money, tore up the royalty contract, effectively walking away from millions in royalties that he was already owed and billions that would have accrued in the future. The Westinghouse Company would continue.

But JP Morgan wasn't done yet. In a move that predicted some of today's most egregious abuses of the patent system, Morgan filed an infringement lawsuit for the Tesla patents that Westinghouse owned. Whatever the legitimacy of the claim, the threat was in the litigation itself: a protracted legal battle would carry expenses ringing into the millions—millions that Morgan, with his diversified empire, could afford to spend, but which Westinghouse could not. In the end, with no resources to fight and no other option, Westinghouse was forced to sign over the Tesla AC patents to the General Electric Company. The "War of the Currents" had been won by Tesla, Westinghouse, and AC power... but the spoils went to JP Morgan.

—

While it is debatable whether JP Morgan was the original gangsta' or merely a product of his era, there's no question that moves like his hostile takeover of the Tesla/Westinghouse patents were emblematic of business practices in the "Gilded Age." Want proof? Here are just a few examples of familiar inventions that have been similarly stolen in the years directly preceding and since:

Monopoly—The famous board game was created by a bold and progressive woman by the name of Elizabeth Magie, in 1903. Back then it was called "The Landlord's Game" and was designed to demonstrate the tragic consequences of land accumulation. Despite the fact that this story is now well known, many still claim that game designer Charles Darrow is the legitimate inventor—even though he simply stamped his name on something that did not belong to him.

The Sewing Machine—When you think of a sewing machine, the first company that probably comes to mind is the Singer Corporation, mainly because they are still a powerful entity in the industry to this day. However, according to Cambridge History,[52] despite this firm control over the sewing machine industry, Isaac Singer and Singer Corporation stole the idea from Elias Howe, who eventually sued the company for the right to receive royalties (fortunately, in this case, the inventor won).

The Television—Although your textbooks may tell you that the first television set was created by Vladimir Zworykin for the RCA electronics company, it was actually invented by Philo Taylor Farnsworth, an inventor with 165 patents to his name. It turned out that Farnsworth invented the television in 1927, at the age of 21; three years later Zworykin visited his laboratory to see his invention... and steal his ideas. After a decade-long court battle, RCA eventually lost their initial case and appeal, meaning that Farnsworth would receive royalties for the inventions, though history is yet to confer on him the full recognition he deserves.

The Telephone—In the late 1800s the race was on to create the first suc-

---

[52] "Elias Howe's Sewing Machine: Main Street near Windsor." https://historycambridge.org/innovation/Sewing%20Machine.html

cessful telephone, and the main contenders in this race were Elisha Gray and Alexander Graham Bell. If you've never heard of Gray, it's probably because you were taught at school that Bell was the genius who invented the device that could transmit intelligible sounds from one place to another. As it happens, on February 14, 1876, both men filed their patents... though it was later discovered that Bell had bribed the patent office to show him what Gray's invention looked like. Because of this deception, it is often claimed that Elisha Gray is the main inventor of the telephone, although he never received a dime for his invention; nor has he received the credit he deserves. When Gray died in 1901, his obituary in the *New York Times* predicted he would "receive full justice at the hands of future historians by being immortalized as the inventor of the speaking telephone." This, unfortunately, has not yet happened.[53]

The Radio—In the last decade of the 19th Century, our old friend Nikola Tesla discovered that he could use his electronically-charged Tesla coils to transmit messages over long distances. He applied for and received a patent for this process in 1900. Around this same time, a young inventor by the name of Marconi was working on something similar. Using unlicensed IP expressly protected by Tesla's patents, Marconi succeeded in creating a radio broadcast, and was hailed as the inventor of the radio. Tesla, unsurprisingly, was furious, though he never had the money to prosecute Marconi for his theft. Thankfully, the invention has since been rightfully credited to Tesla—though this only occurred after his death.

Jack Daniel's—This Tennessee whiskey distillery claimed that, while the slaves helped create the recipe, they had not really worked out the exact process that made Jack Daniel's whiskey so delicious: that honor, they claimed, belonged to the founder, Jack Daniel. However, in 2016, the company that owns the Jack Daniel's distillery, Brown-Foreman, made headlines when they finally decided to give recognition to the slave who had created the whiskey, Nearest Green.

---

[53] "Elisha Gray and the Telephone: Does This Ring a Bell?" Graybar Electric Company Inc. https://poweringthenewera.com/elisha-gray-and-the-telephone-does-this-ring-a-bell/

The Light Bulb—While Thomas Edison was a brilliant inventor, it now seems clear that he did not, in fact, invent the light bulb. That honor goes to an inventor named Heinrich Goebel, who was reported to have created an incandescent lightbulb 25 years before Edison filed his patent application. In fact, it is reported that at one point Goebel even tried to sell the device to Edison; Edison, apparently, did not see anything useful in it at the time, and refused the offer.

III. THE MODERN AGE, THE BIRTH OF DIGITAL, AND THE DEATH OF IP

Despite the contention that seemed to continuously orbit innovation and IP during the mad rush of the Second Industrial Revolution, patent legislation itself remained largely unchanged; it wasn't until 1952, in fact, that any substantive revisions were made at all. The Patent Act of 1952 included for the first time a requirement that, in order to be patentable, an invention had to be not only novel but also had to include a definition of infringement—a definition and distinction previously left to the courts.[54] Even with these changes, though, the patent system would continue to work in much the same way as it had since the Patent Act of 1836... until 1994, that is, when things started to go sideways.

In the late 1990s Bill Gates, probably feeling pretty high and mighty as the founder of Microsoft and, at the time, the richest man in the world, was asked by an interviewer which of his competitors kept him awake at night. Was it AOL? Intel? Apple? Gates replied that it wasn't any of these: what he worried

---

[54] "A Brief History of the Patent Law of the United States," Ladas & Parry. https://ladas.-com/education-center/a-brief-history-of-the-patent-law-of-the-united-states-2/

about, he said, was a couple guys in a garage, hard at work on the next big thing.

We have to imagine that Gates' concern didn't start in '98 with the interviewer's question, but stretched back to the beginning of his own rise to technological supremacy. Likewise we have to imagine he wasn't the only one who felt this level of concern for the unknown and unanticipated dark horse inventor coming out of nowhere to eat the emerging tech moguls' lunch.

So what did they do with all of their concern and all of their fortune and influence? Let me draw a couple of dots for you, and we'll see if we can't work together to connect them.

In 1994, at the dawn of the world wide web (according to some estimates there were just 10,000 websites and 2,000,000 computers connected to the internet in 1994[55]), while Amazon, Yahoo!, and Mosaic Communications were all in their beginning stages and the first commercial web browser—Netscape Navigator—was still in its after-launch infancy, while AOL and CompuServe were helping home users dial up a connection and the first online transaction (allegedly for a pizza from Pizza Hut) was taking place, the United States Patent and Trademark Office (USPTO) was quietly establishing a program that would fundamentally alter the patent application process, placing a thumb firmly on the scales in favor of the emerging Big Tech interests.

The existence of the Sensitive Application Warning System (SAWS) program wasn't made public until 2015, and relatively little is still known about its inner workings: like a black hole, most of what is known is gleaned by observing the effect it has on the objects that have been unlucky enough to fall into its sphere of influence. However, at this point, enough has been discovered and inferred to draw the charge that the program—which presents itself as a gesture toward diligence and due process, and a check on the quality of the examinations be-

---

[55] "1994 in technology: What the Internet, computers and phones were like 20 years ago," Geoff Herbert, Syracuse.com, March 22, 2019. https://www.syracuse.com/news/2014/11/technology_history_internet_computers_phones_1994.html

ing done—is in fact little more than a means to waylay certain applications in a never-ending examination process, ensuring that they are never, ever approved.

Why would such a program exist? Who would benefit from it? Whose interests would it serve? Devon Rolf, an inventor and patent lawyer, sheds some light on the subject in a piece he published in IPWatchdog.com.[56] Essentially, Devon and his company, Gofigure, created iTunes before Apple even came up with iTunes; he got there first, but... I'll let Devon tell you the rest.

"I first became aware of the USPTO's Sensitive Application Warning System (SAWS) policy in January 2008 when a patent examiner told me during a phone call that he had just been instructed not to allow patent applications that he was examining and that were related to Gofigure's U.S. Patent No. 7,065,342, entitled 'System and Mobile Cellular Telephone Device for Playing Recorded Music.' The '342 patent, which has a priority date in 1999, relates generally to a smartphone for downloading and streaming music and storing the title of a purchased music recording in an account of the user. The examiner said that Gofigure's patent applications related to the '342 patent were in the Sensitive Application Warning System, or SAWS. Neither I (nor Gofigure's patent counsel) had heard of SAWS."

Devon and his team did some digging. According to Devon's article: "through the USPTO's Ombudsman program, Gofigure's patent counsel was told by a USPTO employee with direct knowledge of Gofigure's patent applications that—despite the fact that the patent application being examined had a priority filing date in 1999, several years prior to the introduction of Apple's iTunes Music Store, the GoFigure application was in the SAWS program because it 'reads on iTunes,' and because granting the GoFigure patent could result in 'a very, very public case,' so the USPTO had to be careful."

Gofigure filed a Freedom of Information Act (FOIA) request to try to learn

---

[56] "Secret Examination Procedures at the USPTO: My Experience with SAWS," Devon Rolf, IP Watchdog, December 14, 2014. https://www.ipwatchdog.com/2014/12/14/secret-examination-procedures-at-the-uspto-my-experience-with-saws/id=52638/

more about SAWS, but the USPTO denied the request: they said that the information shared would pose a serious risk to the USPTO's screening process under this policy, and that the USPTO was not required to disclose internal deliberations by USPTO employees. Gofigure appealed the decision, but this appeal was denied.

Devon had created and filed a patent application for a system for downloading, storing, and cataloging digital music before iTunes even existed, but somewhere along the line, between application and approval, Apple stepped into the game. How do we know it was after? Because Devon's application wasn't rejected for getting to the ball second, as it would rightfully have been if that was the case. Instead, it was simply hung up somewhere between third and home, left in a perpetual holding pattern until finally being spit out the other end, rejected for being in SAWs in the first place. Devon explains: "the patent examiner told our patent counsel that he would allow one of Gofigure's patent applications if certain claim amendments were made. We agreed to make the amendments. However, a week later, the patent examiner called Gofigure's patent counsel to explain that he was 'sorry' to inform us that the application was in SAWS and therefore 'cannot be allowed—that is the rule.' The examiner [then] stated that he had to reject the application... [he] stated that, when he had tried to allow the patent application, the USPTO system returned a thread—'SAWS case—cannot be allowed.' The application was indeed rejected."[57]

Though the agency officially retired the SAWS program in 2015,[58] the full story is far less encouraging. Conveniently concurrent with the official end of SAWS, the USPTO launched a new "Enhanced Patent Quality Initiative" (EPQI) encompassing a dozen programs and designed to provide "high-quality, effi-

---

[57] *Ibid.*

[58] "SAWS Retired by USPTO," Gene Quinn, IP Watchdog, March 2, 2015. https://www.ip-watchdog.com/2015/03/02/saws-retired-by-uspto/id=55329/

cient examination of patent applications."[59] Critics and watchdog groups charge that this new EPQI is quietly continuing the work, picking up right where SAWS left off without even missing a beat.

So now we've got our dots—Big Tech (with Big Tech money) on the one side, small company with an inconvenient patent application on the other, and a government agency in the middle—can you see the picture they're forming yet? Look close. I'll give you a hint: it's a big ol' middle finger, and it's pointed right at you, the indie inventor.

False victory or not, by 2015 indie inventors had a bigger problem than the "greatly exaggerated" reports of the SAWS program's demise. By then Big Tech had pulled off a much bigger coup. You see, as the first decade of the new century drew to a close those Big Tech big dogs decided to do ol' JP Morgan one better, and set their sights on the United States patent system itself.

A strong patent system means an inventor can actually monetize his or her invention, attract investment, justify research and development efforts, and develop new markets for new products. A patentee must be confident that his or her issued (and paid for) United States patent will be enforced by the issuing government and therefore respected by competitors. If the patent system is weak in its enforcement of patent rights, the free riders, copiers, and patent thieves can simply take the IP and leave the inventor with nothing for his or her efforts. Under such circumstances, inventors are left with no alternative but to hide their inventions away as trade secrets or risk losing them altogether. Innovations that might have spread far and wide under licensing deals, spurring industry and prompting job growth and further innovation, instead languish in private ownership—the very situation our Founding Fathers sought to avoid in their creation of a strong patent system in the first place.

But just as a weak patent system is poison for the innovators themselves, it is food for large multinationals. The companies already dominating their fields

---

[59] "Patent Quality," USPTO.gov. https://www.uspto.gov/patents/patent-quality

live on market share, not innovation: while an indie inventor needs something truly novel to step into and stand out in the market, Apple needs only to add a third camera lens to its iPhone, or make the device's screen bigger. And yet a Big Tech company like Google remains fundamentally vulnerable to competition: if another search engine company—perhaps "a couple guys in a garage"—can create a program that outperforms Google's algorithms, then Google quickly topples from its perch atop the heap. It becomes obsolete and loses its market dominance.

However. The only way this unseating can happen is if the innovation coming out of this garage is protected by a patent that is itself enforced by the government that issued it: without that, Google could simply replicate the new operation and integrate it into itself, or else dip into its hundreds of billions in assets to litigate the company into nonexistence while simultaneously offering to lowball it into assimilation. Unless a patent is held sacrosanct—even against the bullying that money can accomplish—it simply isn't worth the paper it's printed on.

Sound extreme? Unlikely? Does it sound like the far-fetched catastrophizing of a paranoid mind? Then prepare yourself, 'cuz here comes another round...

As the 21st Century beckoned and the world turned digital, "innovation" and "invention" began to reach beyond the limits of what the patent system had been established to handle hundreds of years before. Software applications stretched the boundaries of patentability, while proprietary algorithms simply stood outside all working definitions (one fundamental patent tenet holds true: you can't patent math). The USPTO wasn't set up to handle a question like, "How much innovation is needed in a line of code for it to constitute a new idea?" In response to the new challenges presented by the ever-increasing complexity of the ever-increasing digital sector, in 2003-2004, the Federal Trade Commission and National Academy of Sciences commissioned and published a set of reports with proposals and recommendations aimed at meeting these challenges and striking a balance between patents and competition in the modern era. Drawing on this report, the proposed Patent Reform Act of 2005 addressed these concerns with, among other things, the proposal that reduced fines for infringement would incentivize patent holders and their competitors to work to productively expand the store of shared patented information, rather than refuse to read each other's patents (this out of fear that it would make them liable to a later accusation of infringement). This Act was not enacted before the 109th Congress concluded, but much of its content was carried forward into the proposed Patent Reform Act of 2007. Significantly this Act also proposed that the American patent system switch from a first-to-invent to a first-to-file system, simplifying the examination process and making the U.S. system more similar to the patent systems already operating in many other modern nations. This bill passed the house but died in the Senate, though much of its content (and, accordingly, the content of the proposed 2005 Act) was carried forward into the Patent Reform Act of 2009; the passage of that bill meant that, finally, the American patent system was catching up to the changes already made by the rest of the world to meet the unique challenges presented by IP in this new digital age.

But these changes were nothing compared to what would come next. As I said, JP Morgan wouldn't be the last to leverage his existing wealth and influence for his own further gain. Obama might have ridden a promise of "hope and change" into the White House in 2008, but I'm pretty sure the change we indie inventors got was not what any of us were looking for.

In 2018, author and IP lawyer Michael Shore published an article on IP-Watchdog.com outlining his take on the patent system in its present state. Titled "How Google and Big Tech Killed the U.S. Patent System,"[60] the article describes the situation this way: "Banana Republics are societies characterized by

---

[60] "How Google and Big Tech Killed the U.S. Patent System," Michael Shore, IP Watchdog, March 2018. https://www.ipwatchdog.com/2018/03/21/how-google-and-big-tech-killed-the-u-s-patent-system/id=95080/

their starkly stratified social classes and a ruling-class plutocracy composed of the business, political and military elites. The Elites rule over a servile government that abets and supports, for kickbacks and bribes, the exploitation of the rest of society. Instead of Dole and United Fruit controlling Honduras, we now have Apple, Microsoft, Amazon, Google and other tech giants controlling Congress and the Executive Branch through unlimited lobbying by groups like the Internet Association, High Tech Inventors Alliance, the Software Alliance, Unified Patents and through direct political donations... the only difference between Honduras in 1904 and the United States today is that the new bananas are smartphones and the software they contain."

What? Wait, what? What the actual fuck happened? How did we go from a Patent Reform Act promising to bring the U.S. system into the modern age—promising to make our patent system more robust in the face of the unique challenges of the digital era—and end up with a situation fitting Shore's description? Where—if we're going to get personal, here—'lil ol' Kip Azzoni Doyle can find herself on the *receiving* end of a lawsuit out of never-even-been-there-Texas, and all for demanding that an infringer quit infringing (read: stealing) from me? Where, to paraphrase my chapter-helper buddy M. David Hoyle, the thief can break into your house and steal everything, and when you catch him and bring him to trial, the judge gives him your house?

You know the drill: you don't remember the last two (or was it three?) shots, and now you're waking up on the lawn with scuffed knees and a cracked phone screen... no real memory of how you got there, just the firm conviction this is *not* where you planned on ending up when you left that house earlier that evening. This is decidedly *not* where you thought this night was headed. Come on. Let's get you up. We can piece it back together. It won't fix the angry mosh pit roiling in your skull, won't un-tear your favorite jeans or un-scuff your knees, but what else can you do? You've got to know what happened, even if knowing doesn't make it any better. Even if it feels like knowing only makes it worse...

Still with me? I knew you would be. Indie-inventor-types don't rattle easily. And anyway: in for a penny in for a pound, amIright? If it was about money or fame we all would have jumped ship a long time ago. What's that phrase? "Throwing good money after bad?" Tattoo it on my forehead, next to the Card-Shark logo and the list of lawyers I've gone through. You know me. We indies are the ones who ride the thing all the way down. And don't worry: we're almost there. Toes keep reaching for the muddy bottom, and pay no mind to the dark shapes you see swimming around in the murk... remember, they're more afraid of you than you are of them. That's why they stacked the deck against you, to kill you where you stand.

You remember that famous scene in *The Matrix*, the one with the red pill and the blue pill? The one where Morpheus explains to Neo how it was the humans' "scorching" the sky, thinking that this would starve the solar-dependent robots, that set in motion the whole terrible chain of events that led to their current predicament? Unfortunately, it's a lot like that. The sorry state of the patent world can be traced back to a reasonable-seeming plan to solve a very real problem. You see, unlike the inventions of the prior era, a single soft- and hardware-based product might contain hundreds—if not thousands—of items of IP: one big tangle of patented and patent-adjacent ideas and methods building on itself and each other. As Tibi Puiu put it in an article published on ZMEScience.com: "Today, your smartphone, pocket calculator, or even a USB-C charger is millions of times more powerful than the Apollo 11 guidance computers."[61] Exciting, yes, but let's take a look at what that really means. A patent represents one idea or claim. A claim is a specific explanation of the parameters of the patented idea. A patent may have a bunch of claims attached to it. In the case of my very finite CardShark, the claims attached to the patent deal with whether a pocket is smooth or rough in nature, or the opening is a slit or a

---

[61] "Your smartphone is millions of times more powerful than the Apollo 11 guidance computers," Tibi Puiu, ZME Science, May 13, 2021. https://www.zmescience.com/science/news-science/smartphone-power-compared-to-apollo-432/

hinged door on the back of the case. Simple but important: the claims set the limit of my patents. Imagine, then, the systems operating in products put out by companies like Apple, Microsoft, Qualcomm, and the other Silicon Valley royalty, each incorporating hundreds if not thousands if not hundreds of thousands of patents, and each of those patents containing dozens if not hundreds if not thousands of claims… and you start to see that you are looking at an absolute tsunami of data. Within this context it was becoming increasingly difficult to discern where one item of patented IP ended and another began: where "use" ended and "innovation" started, and where the last straw broke the camel's back and a patented idea or practice was altered enough to constitute a new idea… and all of it happening at the speed of broadband. Unsurprisingly, this situation created a cascade of IP-related litigation, and an accordant and unsurprising backlash within the culture of the emerging Tech Giants: after all, the thinking seemed to go, they were the ones bringing America into the future, doing the important innovation work that would fuel the next "moonshot" and keep America vital in the face of surging international competition, and here they were bogged down with legal action from litigants who wanted to argue over whose codes are whose, over who wrote what on a cocktail napkin in some bar five years ago.[62]

All of this might have been one thing and might have been sorted out in a more amicable (and less burn-the-fields-and-salt-the-earth) fashion—might have produced more equitable legislative reform—if not for one thing. The very na-

---

[62] Which is not to say that Big Tech was innocent in all—or in any—of this. Remember my buddy M. David Hoyle? In 1997 and 1998 David's company, B.E. Technologies, held patents pertaining to targeted advertising. According to B.E.'s website at the time, it was their intention to be "the dominant interface between end users and their data, anywhere, anytime and from any device." Flash forward to 2005, and Google has effectively subsumed B.E.'s entire patent platform, even going so far as to lift their mission statement: with Google, "users will have immediate access to valuable information—anywhere, anytime and from any device." No license, no acknowledgement, only the promise of a protracted legal battle if David wanted to take up the issue… one which they—and their deep, deep pockets—could afford to wage. All of this merits further explanation: you'll learn more about David and his battle with his own Goliath in Parts III and IV.

ture of the situation surrounding IP in the digital age created a specific kind of opportunity for a specific kind of third party now coming out of the shadows to rear its ugly head. See, patents aren't just *about* products, they *are* products themselves, and can be bought and sold like any other item in the marketplace. If you have a patent that you think has the potential to make you $10,000 over the course of 10 years, and I come along and offer you $7,500 right now to buy it outright, you might take that deal and take a nice vacation, rather than wait 10 years for your ship to come in. I could then take your (now, my) patent and produce your patented product or idea, or I could license it to someone else, or I could look around for anyone I think might be infringing on your (read: MY) patent... and sue the living shit out of them.

"Patent trolling" and the "patent trolls" who practice it: vocab words I could never have imagined I would learn, and wish I'd never had reason to. Turns out, though, they—or at least the boogeyman threat of them—have everything to do with how we got here. As I've said, the tangled nature of the IP coming out of Silicon Valley was making litigation a fact of life for Big Tech. At the same time, however, they were swimming in capital, and inclined to settle any dispute just to keep themselves and their resources from getting hung up in protracted court proceedings. Under these circumstances, a motivated third party could buy up patents pertaining to tech and then sue each and every Big Tech entity whose products' DNA *juuuuuusssssttt miiiiiiggghhhttt* overlap with that of their newly-purchased patented IP. And, to paraphrase Edward Norton's character in *Fight Club*: if the cost to settle is less than the cost to battle it out in court, then the defendant doesn't bother. They settle out-of-pocket and the troll walks away with a nice payday for his troubles. And there it is: your four-step guide to being a Patent Troll. Purchase, sue, settle, repeat: it's fun and easy! Never mind that, to stop you and your Troll cohorts, the human race might get desperate enough to scorch the sky...

The America Invents Act was signed into law in 2011 and took effect in 2012. President Obama called it a "long overdue reform, vital to our ongoing efforts to modernize America's patent laws." What it was, in fact, was Google's Death Star, brought to terrible, fully operational life at the hands of a Congress that had been hand-led down the garden path by Google's rapidly emerging and expanding influence machine.

Stunned? Confused? Don't worry, babyboo. I got you. I'll explain. In order to do it, though, I'm going to lean pretty heavily on David Dayen and his 2016 article "The Android Administration,"[63] published in *The Intercept*:

Google "paid almost no attention to the Washington influence game prior to 2007, but ramped up [its efforts] steeply thereafter." Its "lobbying strategy... include[d] throwing lavish D.C. parties; making grants to trade groups, advocacy organizations, and think tanks; offering free services and training to campaigns, congressional offices, and journalists; and using academics as validators for the company's public policy positions.* Eric Schmidt, executive chairman of Alphabet, Google's parent company, was an enthusiastic supporter of both of Obama's presidential campaigns and has been a major Democratic donor." And the love didn't flow just one way: "For its part, the Obama administration—attempting to project a brand of innovative, post-partisan problem-solving of issues that [had] bedeviled government for decades—welcomed and even [came] to depend upon its association with one of America's largest tech companies."

*There is so, sooooo much to go into about Google and its various interventions in and manipulations of the governmental policies attempting to affect it (a book 10 times as long as this one would just scratch the surface)— much more, notably, than fits into a standard footnote, hence this deviation from form. I do, however, want to take a moment aside to tell you a little bit about this one particular item: it is, to me, one of the most stunning examples of the lengths to which Google will go to skew the odds in their favor. According to a July 2017 Wall Street Journal article by Brody Mullins and Jack Nicas

---

[63] "The Android Administration," David Dayen, The Intercept, April 22 2016. https://theintercept.com/2016/04/22/googles-remarkably-close-relationship-with-the-obama-white-house-in-two-charts/

and titled "Hidden Influence | Paying Professors: Inside Google's Academic Influence Campaign," Google has long operated a "little-known program to harness the brain power of university researchers to help sway opinion and public policy, cultivating financial relationships with professors at campuses from Harvard University to the University of California, Berkeley" and financing "hundreds of research papers to defend against regulatory challenges of its market dominance, paying $5,000 to $400,000 for the work."

According to a concurrent examination by the Tech Transparency Project ("a research initiative of Campaign for Accountability, a 501(c)(3) non-profit, non-partisan watchdog organization that uses research, litigation and aggressive communications to expose how decisions made behind the doors of corporate boardrooms and government offices impact Americans' lives"), between 2005 and 2017, Google exercised an "increasingly pernicious" influence on the academic research happening within the silo of its concern, paying "millions of dollars each year to academics and scholars who produce papers that support its business and policy goals." In all, the Project "identified 330 research papers published between 2005 and 2017 on public policy matters of interest to Google that were in some way funded by the company." According to a former lobbyist, Google "promot[ed] [these] research papers to government officials, and sometimes paid travel expenses for professors to meet with congressional aides and administration officials" in an overt effort "to influence decision makers... on public-policy matters." According to the Tech Transparency Project report, "readers of the papers would not have been aware of the corporate funding: Academics did not disclose the Google funding in nearly two-thirds of cases (65%)," and "[a]uthors failed to disclose funding even when they were directly funded by Google in more than a quarter (26%) of cases." According to the Report, these academic papers "encompassed a wide range of policy and legal issues of critical importance to Google's bottom line, including antitrust, privacy, net neutrality, search neutrality, patents and copyright. They were also tied to specific issues that Google sought to influence."

77

From the Report:

The Google-funded studies came from a wide variety of sources, and often blurred the line between academic research and paid advocacy by the company's consultants. They were authored by academics, think-tanks, law firms, and economic consultants from some of the leading law schools and universities in the country, including Stanford, Harvard, MIT, University of California Berkeley, UCLA, Rutgers, Georgetown, Northwestern Law School, and Columbia...

Many of the Google-funded policy research papers examined were not published in peer-reviewed journals. Some were self-published on the Social Science Research Network, and many more appeared in publications that lack peer-review requirements. Our analysis showed that Google-funded studies routinely cited each other. The practice helps obscure the original Google funding and creates the impression of a large and growing body of academic research that supports the company's policy positions....

Google's paid policy research had broad reach and may have influenced policymakers unaware of its sponsorship. Google lobbyists and lawyers pushed the Google-funded research to journalists, the White House, Congress, regulators and agencies investigating its conduct, such as the Federal Trade Commission, often without disclosing that they paid for its production...

Overall, our analysis suggests that Google is using its sponsorship of academic research, not to advance knowledge and understanding, but as an extension of its public relations and influence machine.[64]

---

[64] See "Hidden Influence | Paying Professors: Inside Google's Academic Influence Campaign," by Brody Mullins and Jack Nicas, the Wall Street Journal, July 14, 2017. https://www.wsj.com/articles/paying-professors-inside-googles-academic-influence-campaign-1499785286, and "Google Academics Inc.," Tech Transparency Project, July 11, 2017. https://www.techtransparencyproject.org/articles/google-academics-inc

Further, in an era defined by ever-more complex technological challenges and demands, Google offered itself as—and functionally became—"a sort of corporate extension of government operations in the digital era."[65] Again, according to David Dayen: "Modern life requires so much information technology support that a sprawling operation like the White House has turned to tech companies... when faced with pressing IT needs. Practically every part of the government makes available some form of technology, whether it's the public-facing website for a federal agency, a digital mechanism for people to access benefits, or a new communications tool for espionage or war. Somebody has to build and manage those projects, and Silicon Valley firms have the expertise needed to do that. White House officials have publicly asked Silicon Valley for aid in stopping terrorists from recruiting via social media, securing the internet of things, thwarting cyberattacks, modernizing the Defense Department, and generally updating all their technology... Google provided diplomatic assistance to the administration through expanding internet access in Cuba; collaborated with the Department of Housing and Urban Development to bring Google Fiber into public housing; used Google resources to monitor droughts in real time; and even captured 360-degree views of White House interiors."[66]

It's the ol' capitalism-democracy two-step: societies have needs, governments exist in part to meet and manage those needs, and—in America—private companies step up to answer the call the government puts out for help. But private companies aren't public utilities: ethical business practices may still exist, but if the last two plus decades of often blatant corporate malfeasance is any indication, corporate loyalty is to itself and its shareholders first and the rest of us a distant second. And, accordingly, in a system in which a private interest can end up partnering with the very government that regulates the poli-

---

[65] "The Android Administration," David Dayen, The Intercept, April 22 2016. https://theintercept.com/2016/04/22/googles-remarkably-close-relationship-with-the-obama-white-house-in-two-charts/

[66] *Ibid.*

cies impacting that interest's bottom line, things can start to go wrong... with grave results for the rest of us. Again, David Dayen:

"According to an analysis of White House data, the Google lobbyist with the most White House visits, Johanna Shelton, visited 128 times, far more often than lead representatives of the other top-lobbying companies—and more than twice as often, for instance, as Microsoft's Fred Humphries or Comcast's David Cohen. Asked to respond, Google spokesperson Riva Litman referred *The Intercept* to a blog post written when the Wall Street Journal raised similar questions a year ago. In that post, Google said the meetings covered a host of topics, including patent reform, STEM education, internet censorship, cloud computing, trade and investment, and smart contact lenses."

...um, what? Mic check one two? Are you hearing me? Is this thing on? Google lobbyist Johanna Shelton visited the White House to discuss *PATENT FUCKING REFORM*?!?! They didn't listen to her, right? They considered the biases and interests of source, right? They took it all with a grain of salt, right? Right? *Right????*

Sadly, no—but you already knew that. Google and its Big Tech cohorts had hatched a plan, and Johanna's sit-downs were just one part of a full-court press. A veritable army of lobbyists had descended on Obama's Congress, and the story they told was compelling. Here they were, poor little Google, working hard to keep America at the head of the pack in the mad race for global tech primacy. They were helping the people's government in ways that only they could. Not only that, they were also donating millions of dollars to party efforts and individual campaigns. But they were playing hurt, hamstrung by endless litigation from these damn pesky patent trolls. If only the patent trolls would go away, they could do so much more. If only there were policies in place that allowed companies to function and thrive in this new digital world! They'd done so much—and would do so much more—for those in the power; wasn't there anything they, the legislators in the position to make the call, could do? And hey, since you're asking, we do have a few suggestions...

You see where this is going, don't you? The America Invents Act—Obama's "vital" and "long overdue reform"—enacted into law a set of significant changes to the patent system that fundamentally eroded patent protection in this country, flying in the face of the spirit of the patent system itself—changes suggested, pushed, bought, and paid for by Google and the other Big Tech companies that its changes benefit. Specifically and perhaps most importantly, it rewrote the book on how infringement cases are handled—and who's doing the deciding when they are.

Prior to the AIA, if you as a patent holder brought a case against someone infringing your IP, and that infringer responded by calling the validity of your patent into question, your case would be heard before the impartial judges of the Article III District Court. Nominated by the president and confirmed by the U.S. Senate, and as outlined in Article III, Section 1 of the Constitution, these judges "shall hold their offices during good behavior, and shall, at stated times, receive for their services, a compensation, which shall not be diminished during their continuance in office." According to law professors Richard W. Garnett and David A. Strauss, this provision "is designed to make sure that the judges are independent. They can decide cases according to what they think the law requires, without worrying about whether some powerful person—or even a majority of the people—will object. As Alexander Hamilton put it in The Federalist No. 78, judicial independence 'is the best expedient which can be devised in any government to secure a steady, upright, and impartial administration of the laws.'"[67] Meaning that, if the validity of your patent was called into question, you were guaranteed a fair hearing before the ruling governing body.

The AIA changed all that. Under the AIA, if an infringer wanted to call the validity of the patent they were infringing into question, they could submit that patent for an "inter partes review" (IPR) before the newly-created Patent Trial and Appeal Board (PTAB). The inter partes review process effectively removes the patent validity trial from the conventional court setting and places it before

---

[67] "Common Interpretation: Article III, Section One," Richard W. Garnett and David A. Strauss, ConstitutionCenter.org. https://constitutioncenter.org/interactive-constitution/interpretation/article-iii/clauses/45

this newly created board—a board made up of appointees with no legal obligation to judicial impartiality, hired and paid by the USPTO Director.

Under the AIA, this tribunal has the unilateral power to render patents invalid.

There's more. Regardless of its outcome, the simple fact of an IPR is enough to render the patent itself a lame duck asset for the duration of the proceeding. After all: how do you attract investors with a hub IP asset whose future is uncertain? Further, any pending infringement litigation or settlement proceedings can be and will be stayed until the IPR resolves, meaning that the patent is *effectively and functionally* invalidated as soon as the IPR goes into effect. This process burns years off the patent's enforceable lifespan and, even if the IPR is eventually struck down, this lost time is not added back to the patent's very limited term. Historically, in many cases, a third or more of the active life of a patent subjected to an IPR is lost. This, despite the fact that the inventor has likely spent years—potentially decades—along with tens of thousands of dollars *proving* to the USPTO that the invention is worthy of a valid patent. Once a PTAB review is instituted, all of this means nothing, and the inventor must go through the time, expense, and effort of proving the validity of their patent all over again—must prove *again* to the very same governmental body that issued their recognition of that patent's validity in the first place, and placed upon it the seal and promise of the United States government.

It gets worse still. The process is expensive: the IPR'd company must pay fees to the PTAB for their case to be heard; this on top of fees paid to lawyers who have to spend the time it takes to become experts on the patent itself and defend it against the company bringing the IPR. And while it is expensive on the other side as well—it can cost upwards of $150,000 to IPR an inventor and his or her idea—a big, deep-pocketed company like Google can pay the money for the IPR knowing that the inventor doesn't have the money to fight as long and as hard as they do. Just like ol' JP Morgan, they can bluster and threaten until the inventor, backed into a financial corner, caves to the pressure. The infringing company keeps infringing and the inventor, with no logistically viable way to fight for what's his or hers, walks away—potentially leaving millions in licensing fees on the table.

Today, under the AIA, any patent can be challenged in front of the PTAB by anyone at any time. Inventors cannot even adjust their own behavior to reduce the risk of an IPR: that risk remains undiminished so long as their patent is in effect. In fact, since most PTAB procedures are launched against patents being enforced against large multinational infringers, the only behavioral change an inventor could make would be to not enforce their patent rights at all: even if a large multinational corporation steals their invention outright. Maybe this, after all, was the point.

Can you feel it? Lungs about to burst, toes reaching down and down for contact that just never comes, and that little ball of light way up there past the surface is starting to look so far away… Hold on a little longer. We've still got farther down to go…

On an administrative level, the PTAB is made up of the four statutory members: the Director of the USPTO, the Deputy Director of the USPTO, the Commissioner for Patents, and the Commissioner for Trademarks. On the ground level, though, actually doing the reviewing, are the administrative patent judges (APJs)—"persons of competent legal knowledge and scientific ability"[68]—three of whom oversee each IPR and rule on the validity of the patent under review. We will discuss these APJs—and the questionable "performance-based" bonus system they operate within—in greater detail in Part IV; for now, though, consider the following. Before the advent of the AIA, Federal District Courts would invalidate a contested patent 28.76% of the time. Under the new system, "62% of completed trials have cancelled all challenged patent claims; and 80% of completed trials have invalidated at least one claim."[69] These are patents and

---

[68] "35 U.S. Code § 6 - Patent Trial and Appeal Board," https://www.law.cornell.edu/us-code/text/35/6

[69] "Financial Incentive Structure for AIA Trials Destroys Due Process at PTAB, New Vision Gaming Argues," Steve Brachmann, IP Watchdog, July 15, 2020. https://www.ipwatch-dog.com/2020/07/15/financial-incentive-structure-aia-trials-destroys-due-process-ptab-new-vision-gaming-argues/id=123303/

patent claims, it must be remembered, that were previously and thoroughly reviewed, deemed valid, and sealed by the U.S. government itself. With such a wide margin between the two systems, one is forced to wonder whether many bad patents escaped the old system unscathed... or whether the PTAB is simply, as many have called it, a designated "death squad killing property rights."[70]

Worse than this, should a patent escape an IPR intact, neither the AIA nor the USPTO rules set a limit on the number of times a patent can be subjected to an IPR. A successful patent defense before the PTAB sets no precedent. Should a given IPR fail to invalidate a patent or claim, a large and deep-pocketed infringer can simply file another one, directly or through surrogates, again and again, until they've achieved the result they're hoping for.[71] No matter how many times this process repeats, or how many times a patent owner wins at the PTAB, their patent is never safe. Surviving one IPR provides no security from future challenges.

It should also be noted that, as there is no standing ethical mandate disallowing it, APJs can sit in consideration of cases involving former employers or clients[72]; they can also rule in favor of a company and then accept a job working at that same company. I would comment on that, but the problems such a

---

[70] This comment was made by Federal Circuit Chief Judge Randall Rader to at the AIPLA annual meeting in October, 2013.

[71] Which is to say nothing of the process of "stacking" panels, or the motivations and machinations of the person assigning the APJs to the case... but we're getting ahead of ourselves. We'll talk more about that in Part IV, when we sit down for a meet and greet with former USPTO Director—and former high-level Google employee—Michelle Lee.

[72] See: "If PTAB judges can decide cases involving former defense clients USPTO conflict rules must change," Gene Quinn, IP Watchdog, May 2, 2017. http://www.ipwatchdog.-com/2017/05/02/ptab-judge-former-clients-uspto-conflict-rules/id=82765/; "Is the ethical bar for practitioners higher than it is for PTAB judges?" Gene Quinn, IP Watchdog, May 3, 2017. http://www.ipwatchdog.com/2017/05/03/ethical-bar-practitioners-higher-ptab-judges/id=82870/; and "More conflicts of interest surface with second PTAB judge," Gene Quinn & Steve Brachmann, IP Watchdog, May 7, 2017. http://www.ip-watchdog.com/2017/05/07/more-conflicts-interest-surface-second-ptab-judge/id=83012/

situation presents seem so blatant and obvious as to make any comment redundant. Instead, I might simply ask if this seems like a situation conducive to the "impartial administration of the laws"; and, if it does not, why a system with such a guarantee in place would ever be replaced by one with no such guarantee... and who such a change might reasonably be expected to benefit.

Perhaps the most insidious damage resulting from this change is the significantly deteriorated value of the patent asset as it relates to an inventor's efforts to secure funding and investment to develop his or her product and bring it to market. Under the prior system, a patent—a transferable private property right—might attract investment at the earliest, high-risk stages of commercialization by virtue of its inherent value. Now an investor is given significant cause for hesitation when considering a company whose viability depends on a patent, as there is no longer any reasonable guarantee that the company won't be IPR'd and stripped of that patent protection. Signing on to fund a company now means, in essence, signing on to defend that company (and your investment) against an IPR—a process which may cost hundreds of thousands of dollars, if not more, and tie both company and investor up in litigation for years. The perverse reality of the AIA is that now, in fact, it is wiser to invest in the company that steals the patented IP than it is to invest in the company that invented it.

And you remember that "cooling effect" our Founding Fathers were so worried about? As we discussed in the introduction, since the introduction of the PTAB and the IPR under the AIA, the U.S. has fallen from first to 12th in the global ranking of its intellectual property systems: we are now tied with Italy and Hungary, behind every other highly developed economy.[73] And again: while the U.S. has been busy cannibalizing its patent system, others have been bolstering their own: China, in particular, has been actively strengthening its patent system, implementing aggressive infringement remedies like injunctions

---

[73] "U.S. Chamber International IP Index: Sixth Edition," February 2018. http://www.the-globalipcenter.com/wp-content/uploads/2018/02/GIPC_IP_Index_2018.pdf

BLOOD IN THE WATER

and higher damages awards.[74] With the way things have been going, it seems only a matter of time before China, with such pro-patent policies and protections in place, becomes the epicenter of the innovation universe.

And what about our ol' friend Google? At the time the AIA went into effect, Google enjoyed a 67% market share.[75] Today Google's market share stands at nearly 80%.[76] Looks like they got exactly what they paid for.

In the coming pages I will introduce you to just a few of the thousands of indies who've been harmed by the changes enacted under the AIA, as well as some of the key players who helped make the PTAB the "death squad" it is now. I will also show you some of the nasty inner workings that made (and make) it all possible, the underhanded and potentially illegal tactics employed, and show you exactly how and why the AIA is a terminally problematic piece of legislation that needs to be undone. For now, though: feel the seaweed against your ankles? Feel the mud under your toes? Push up and kick hard. Fight the blackness vignetting your vision. You can do it. You're almost there. Here it comes. Break on through. Deep breaths, now. You're all right. I know you were scared. It's OK to cry. But I need you to be brave, now. I need you to be strong. We've got a long way to go before we're safely back on shore, and in these shark-infested waters—as my first lawyer tried to warn me—absolutely nothing is guaranteed.

---

[74] "Patent Reform With Chinese Characteristics," Ronald A. Cass, the Wall Street Journal, February 10, 2009. https://www.wsj.com/articles/SB123419814824764201.

[75] "Google Takes 67% Search Engine Market Share," Miranda Miller, Search Engine Watch, December 17, 2012. https://searchenginewatch.com/sew/news/2232359/google-takes-67-search-engine-market-share.

[76] https://netmarketshare.com/search-engine-market-share.aspx?options=

PART III

BOOK IT

"If I am not for myself, who will be for me?
If I am not for others, what am I?
And if not now, when?"
—Hillel the Elder

If the definition of insanity is smacking your head against the same wall over and over again and hoping for a different outcome, then let's just say that by the summer of 2010 I was bruised like a peach and ready to strap on a helmet and stop the insanity. The 10th-or-so infringer had started selling their version of my patented idea *sans license* and, despite my deploying my lawyers, sending out my cease-and-desist letter—i.e. *exercising my rights as a patent holder*—I wasn't making an inch of headway. Nor had these same efforts had a much greater effect with the nine (or so) previous infringers. Clearly, something wasn't working.

By now, though, I knew it wasn't just me who was being impacted. In my effort to orient myself in the bizarro world I'd stumbled into I'd connected—via online chat rooms—with a loose and motley gang of indie inventors, each with a patent-infringement horror story worse than the next. On these websites and forums I read story after story about indie inventors who'd lost everything, their patents invalidated by IPRs faster than the flush of a toilet. Some had lost their early investors because, with the way things were now, there was simply no reason to risk an investment in a patented idea anymore—not when your return is much more assured when you invest in the company *stealing* that patented idea. Other inventors had watched their product sales shrink and disappear as infringing products—priced strategically lower than the patent-protected original—subsumed their retail shelf space.

What do you do when you see a wrong that needs righting? You get involved! This is a democracy, right? Start a petition! Write to your state representative! Submit testimony to the subcommittee handling your specific issue! That's how things get done, right? That's how you "be the change you want to see in the world," right? Right?

Yeah... but... no. We fucking *did* all that. I and my online compatriots down in the trenches of patent warfare had *done* all of that, and it hadn't moved the needle one iota.[77] Nothing was changing. My head ached from banging it against the issue. Things weren't getting any better, nor did it seem like they were going to start getting better—not for me and my CardShark, anyway. The booming of Facebook and the advent of Instagram, not to mention Amazon, meant that every day a new infringer was selling their infringing products to a global audience of phone users with the single click of a button... and it really seemed like there was nothing I could do about it.

But hey, like I said: if we indie inventor types knew how to take the world as it is—if we knew how to accept things the way they are and just move the fuck on already—and if we knew how to cash out, kick back, and throw up our hands when things got gravelly... well then, we wouldn't be indie inventors, now would we?

---

[77] More on this later... stay tuned. For now, trust me: we've tried and are still trying.

Because here's the thing. I'd lived a few different lives before I had the gas-pump-side stroke of inspiration that led to the CardShark. Before that, I'd spent plenty of time clickety-clacking over the keys of my laptop. Among other things, I had a masters in screenwriting from NYU, had co-authored a book for Simon & Schuster (*Selling Dreams: How to Make Any Product Irresistible*), and had penned screenplays and published magazine articles on everything from makeup products (I had modeled for Wilhelmina in N.Y.C., which apparently made me an authority on the subject)[78] to the epidemic of kidnappings that has paralyzed Venezuela (I embedded with the counter sequester unit) to car racing (specifically, the 24-hour endurance races). If nothing else—if I didn't know how to wave a magic wand and cure the illness that had crept in and taken hold of the patent system—I at least knew how to put the story down on paper, and I had the contacts and connections from my previous literary and journalistic undertakings to get that story—and its dire warning—out to the people who needed to hear it.

"Shiiiit," I said, "I think I need to write a book."

Like with the CardShark, it was a case of a lot of head nods and "yeses" coming back at me. My inner circle was enthusiastic. "Yes!" they all said. "Put it in a book! The world needs to know what's happening to you, and what's happening to the patent system!"

And sure, it's possible that the subtext of all of this support was, "Please just stop telling us about it for like maybe five minutes," but I wasn't hearing that.

---

[78] When I was "discovered" by the Wilhelmina modeling agent I was working on the tourist boats in Boothbay Harbor, Maine. Someone took a polaroid of me filleting a codfish or scrubbing down the remnants of some seasick tourist—real glam stuff—sent it down to N.Y.C., and the next thing you know... I wasn't a particularly good model: going to school at night and modeling during the day, I'd inevitably work one thing off the other. "Ahh hell, I have a big job and I couldn't get my paper done, Professor..." Similarly, I couldn't go to all my "go-sees" 'cuz I "had a big paper to write." The point being that I was never committed to a modeling career—even back then I saw it as a means to an end—though it would fuel the writing of a satire about the modeling world, as yet unpublished, appropriately titled *Bubblegum for Breakfast*.

The world needed to know what was happening and, as far as I could see, I was just the chick to tell them.

Truth be told, I probably should have learned my lesson with the Card-Shark. I jumped into this project with a lot of enthusiasm and faith that, when all was said and done, a solid and useful product would be out there, connecting with its intended audience and adding something to the world. Of course, just like with the CardShark, I had no idea just how much time, sweat, and frustration would go into what seemed like a fairly simple idea. Inventing the CardShark felt like going the distance in one of these 24-hour races I'd covered, and if I'd known I would be in for more of the same—more pre-dawn hours strapped in the seat with stress, fatigue, and the promise of more heartache riding shotgun—I'm not sure I would have gathered up the gumption to keep moving forward. And, honestly, if it was just me—if it was just about Kip and the CardShark and the injustice of it all—I might not have. I might have called it a day, shoved these pages into a drawer, and gone back to soaking up whatever comfort could be gleaned bitching to friends and family about everything that was so very, very wrong with the patent system. But it wasn't just me. I knew too much, now. I'd connected with too many other indie inventors, heard too many stories from inventors who'd done everything right and still lost everything, whose valid patents had been stripped from them, who'd been left with no choice but to walk away when it became clear that the millions of dollars in licenses they were rightfully owed just weren't coming, who'd been forced to walk away from inventions that had been outright stolen, who had finally burned their patents because those patents were no longer worth the paper they were printed on.[79] These people were the true embodiment of the American spirit—that driving, pioneering faith that hard work and personal striving can build a brighter tomorrow for you, your family, and your country—and they were being thrown to the dogs by the system they'd been told would protect

---

[79] This seriously happened; I'll tell you all about it in Part V.

them. They deserved better; they were *owed* better. *We the People* owed them better. And if the bill hadn't yet been delivered to those it was due—to those with the position and the power to set this badly listing ship aright—then I at least could do my part to hand it on up.[80]

The original idea for this book was to collect the stories of indie inventors and show how they'd been screwed by a crumbling system and the corporate bullies whose fat asses were causing it to crumble. I would intersperse these with interviews with experts in the space: lawyers and academics who knew the whys and the wherefores behind what these indie inventors were going through. Along with that I would tell my own story in the hope that, by show-ing my own missteps and mistakes, I might be able to help other would-be in-dies avoid them. As with any invention, though, the further in I got the more I realized that I didn't know what I didn't know about what I was undertaking. It became very clear very quickly that the scope of the project would have to be significantly expanded if I was to have any hope of doing the subject justice, and that even then I would only be scratching the surface, retelling only a handful of the thousands of stories and mapping out only a sketch of the labyrinthian history of questionable legislation, backroom dealings, and inside baseball that got us here. It is my hope, however, that if I do at least this much well, you will be motivated to go out and learn more: to visit the resources I'm going to tell you about in the coming pages, and to get involved with the im-portant work being done there by the true heroes fighting the good fight for America's very soul.

In the next few pages I'm going to tell you about some of the resources I've

---

[80] And also: Scott. As I said in Part I, I'm usually pretty good about knowing when I've paddled up the edge of my knowledge and abilities, and when it came to this project I didn't need any special awareness: I ran into that edge pretty damn hard. My original crack at telling this story needed help *badly*. Luckily, in that moment, fate (a.k.a. my lawyer, Karl Maersch) introduced me to Scott. Scott helped me say what I had to say better than I could say it, and helped me say more than I knew I needed to say; the re-sultant book—the book you're holding in your hands—despite being written from my point of view, is truly a collaborative effort.

used and that have helped me immensely in both my personal and journalistic efforts to get my bearings in and my arms around the upside-down world of the modern patent system. These resources are where I learned many of the stories that will follow: stories of inventors chewed up and spit out by that same system. This is followed by interviews with experts in the field who, hopefully, can tell us what it all means—and what can be done.

—

IP WATCHDOG

Since its inception in 1999, IPWatchdog.com has served as a resource on intellectual property for tens of millions of site visitors. It offers coverage on matters relating to trade secrets, copyrights, and trademarks, and is recognized as one of the leading online sources for news, information, analysis, and commentary in the patent and innovation industries. Currently it is the largest online IP publication in the world. The site's founder, President, CEO, and primary contributor is Gene Quinn, a patent attorney and commentator on patent law and innovation policy. Gene has been named one of the top 50 most influential people in IP by *Managing IP Magazine*, and in 2017 and 2018 was recognized by *IAM Magazine* as one of the top IP strategists in the world. I can't recall exactly how I first learned about this site, but I know it was a result of my being sucked down some "friend-of-a-friend" vortex which finally ended with my picking up the phone and giving Gene a call. He was, from the start, very gracious, and has gone above and beyond in helping me put this book together. His legal alacrity regarding the patent system leaves me floored every time.

If any of you are planning on staying in this IP fight, I suggest you make Gene's site part of your daily vitamin regiment. Gene is absolutely dedicated to us underdogs, and works tirelessly to break down what's going on, what we can expect, and what we need to brace ourselves for. It is Gene Quinn—along with his activist inventor counterparts at U.S. Inventor: Randy Landreneau, Josh Malone, and Paul Morinville—who are leading the charge for indie inventors and trying to mount a campaign to overhaul the gutted patent system.

Looking for a good place to start? Check out the article "Reverse Patent Reform in 2017 or Wipe out a Generation of Inventors."[81] Gene Quinn and Paul Morinville fire off bullet points charting the veritable bullet wounds inflicted on the patent system and the livelihoods of those who depend on it. From describing the functioning of the PTAB to illustrating how the increased costs associated with patent defense have reduced the intrinsic value of the patent itself, the article clearly and succinctly explains how the threat of an IPR has flipped the economics surrounding patents: how investors and contingency patent litigators[82] have fled the space, leaving indie inventors in the wind. They go on to describe how, in the wake of this phenomenon, patent-holding companies—companies that invest in patent portfolios to profit by their licensing, and which once served as a potential customer base for indie patent holders—have shifted their focus elsewhere: what choice do they have when, with the way things are, a small portfolio's contents might be easily IPR'd into non-existence? Bad as all of that is, the authors close with this further warning: "While there isn't much of a patent system left for U.S. inventors and startups, the real dam-

---

[81] "Reverse Patent Reform in 2017 or Wipe out a Generation of Inventors," Paul Morinville & Gene Quinn, IP Watchdog, January 3, 2017. https://www.ipwatchdog.com/2017/01/03/reverse-patent-reform-2017/id=76221/

[82] Contingency litigators work on spec, and get paid when they get you paid. In the IPR-haunted landscape of modern IP, where the likelihood of a lucrative settlement for a patent holder has dwindled down to a shadow of its former self, most law firms operating in the space have made the economic decision to not do business on contingency, removing what was once a vital tool for the cash-strapped infringee.

age is bigger than just that. While our government has been killing imaginary patent trolls, China has been strengthening their own patent system. China now leads the world in new patent filings. A huge percentage of venture capital has moved to China as a result. We are giving our economic engine to China."[83] Feel free to say it with me: Fucking *yikes.*

U.S. INVENTOR

It came down to the flip of a coin, deciding whether to put U.S. Inventor or IP Watchdog first on this list, so don't let the order fool you. Again, I don't remember exactly how I came across this incredible resource, but I do know that this site and its three creators and primary contributors—Randy Landreneau, Josh Malone, and Paul Morinville—are doing the Lord's work in the patent space, raising petitions and pushing legislation designed to help inventors protect their rights. The site's stated mission is to "teach, promote, and defend the invention process and business methods involved in developing an idea, making a profit, and changing lives." Check out the article "The Loss of Inventor Rights – A Concise Description"[84] to get started, and then—once you're all good and pissed off from what you read there—navigate over to their "Action" and "Donate" tabs to get involved.

---

[83] "Reverse Patent Reform in 2017 or Wipe out a Generation of Inventors," Paul Morinville @ Gene Quinn, IP Watchdog, January 3, 2017. https://www.ipwatchdog.com/2017/01/03/reverse-patent-reform-2017/id=76221/

[84] "The Loss of Inventor Rights – A Concise Description," Randy Landreneau, U.S. Inventor, July 13, 2021. https://usinventor.org/the-loss-of-inventor-rights-a-concise-description/

THE COUNTERFEIT REPORT

The weakened patent system doesn't just lead to inventors being ripped off by infringers: disregard for the enforcement of patents coupled with the explosive growth in internet shopping have created a perfect storm for a surge in counterfeit products.

While an infringer steals an idea and builds their product or product line on IP protected in someone else's patent, a counterfeiter does one better. A counterfeiter simply takes a successful product, copycats that product and its packaging, and pretends to be the original.

If imitation is the sincerest form of flattery then I guess, in some twisted way, counterfeiting is high praise. Too bad it's the kind of praise that literally bankrupts those toiling to bring something new to market, creating jobs and opportunities. "We love your product... and your sales. Think we'll take both." Yeah... how 'bout 'cha don't. How 'bout 'cha leave me and the fruits of my labors the *irrumabo*[85] alone. How 'bout 'cha crawl back into whatever hole you crawled out of and stay there until you've learned your lesson.

Thankfully, here to help all of us keep track of the counterfeit racket is The Counterfeit Report, a consumer advocate and watchdog. To date TCR has identified over 27,200 counterfeit items on Amazon and, acting as the IP owner's representative, have reported 14,535 of these products to Amazon for removal (Amazon, with its global marketplace functionality, is proving to be an ideal platform for the distribution of counterfeit goods). This is important work, and it goes beyond mere economics: counterfeit products are often made with inferior materials and in a manner designed to produce visual but not functional equivalency to the original; accordingly, depending on the manner in which the product is intended to be used, they may be dangerous or even deadly (ac-

---

[85] The *fuck* alone. Latin. You guys are gonna ace the Latin exam at the end of this book.

cordingly, consumers are *strongly* advised to buy directly from the manufacturer or an authorized retailer).

Quick hits from the site: The Counterfeit Report is "the first and only website to promote counterfeit awareness and provide consumers a free and informative visual guide to detecting counterfeit products."[86] It "uses thousands of authentic and counterfeit product photos to show consumers the sophistication of counterfeiters and their ability to create visually deceptive counterfeit products and packaging. Manufacturers… can immediately list and update their counterfeit product information for enhanced brand protection and direct consumer education. Consumers can report seeing or purchasing counterfeit products and the source directly to the manufacturer and authorities on the website."[87] Check them out at www.TheCounterfeitReport.com.

—

INDIE INVENTOR LINDA GOMEZ

The Counterfeit Report is how I originally learned about the indie inventor I'm going to profile first, Linda Gomez. Linda's story is emblematic of the core mission behind the Counterfeit Report. I reached out to Linda for an interview,

---

[86] "Infamous for Counterfeits and Flawed Products, Wish.com has Filed a Draft with SEC to go Public," Craig Crosby, Santa Monica Observer, October 14, 2020. https://www.s-mobserved.com/story/2020/10/14/news/infamous-for-counterfeits-and-flawed-products-wishcom-has-filed-a-draft-with-sec-to-go-public/4969.html

[87] "Amazon Flooded with 75,000 Arbitration Requests and then Quietly Drops Arbitration Clause," Craig Crosby, Santa Monica Observer, June 9, 2021. https://www.smobserved.-com/story/2021/06/09/business/amazon-flooded-with-75000-arbitration-requests-and-then-quietly-drops-arbitration-clause/5723.html

and she was good enough to send over the following narrative recounting all that has happened to her and her *Fullips* line of products.[88] As you'll see, for Linda, trying to protect her authentic product from the counterfeiters has been like playing a game of Whac-A-Mole on LSD: it's a bad trip that just… doesn't… end.

Knocking Off the American Dream

How does a small, "Made in the U.S.A." company survive when its own Government cannot protect it against counterfeits—counterfeits that are being brought into our country from China and sold by the thousands on U.S.-based websites Amazon, eBay, and other private U.S. websites? Counterfeits that are being advertised on Google, another U.S. company. How are these counterfeits even getting here? Easy . . . through Alibaba, AliExpress, and their related affiliates.

I am the owner of a small beauty company, originally based out of Paradise Valley, Arizona. My company makes and sells a popular self-suction lip enhancer. In the U.S., our product is patented, our name is trademarked, and our logo is copyrighted. Indeed, we even have a Chinese trademark on our name and a patent on our product in China.

I thought I had done everything "right." I thought I took all the right steps to protect what I was working so hard to build, even though taking those steps required a lot of time, money, and research.

In the midst of the downturn of the economy, many people, including seniors like myself, went out, "pulled ourselves up by the boot straps" and set out to use that good old American ingenuity and make things happen! That's what we were taught to do, it's how we were raised and it's what we know how to do. We take care of business and pay our bills the old-fashioned way—

---

[88] The *Fullips* device is an FDA-approved silicone suction cup designed to (temporarily) give the user Kim-Kardashian-full lips, without the surgery. You can find it at https://www.fullips.com

start a small business, work hard, take care of our families and hopefully along the way, provide jobs for others as well.

That is exactly what I did (in my 50s) after the real estate market, an area that I worked in, collapsed. Upside down on our home, instead of letting it go back to the bank, I came up with an idea for a product, figured out how to patent it, had a prototype made, and began manufacturing this small beauty tool for women.

With the help and support of my family and friends, years of research, and lots of trial and error, I managed to bring my little product to market. I started out very slowly with sales, working hard to get exposure through on-line publications and hiring others to help.

Slowly, things started to happen. My little company was starting to grow and sell products. I got noticed by Cosmopolitan Magazine, *which gave us a "thumbs up" on YouTube. That was followed by Kathie Lee and Hoda having some fun (perhaps even poking fun) with my product on* TODAY. *Then some wonderful beauty bloggers like Michelle Phan picked up my product, helping to promote us on social media. It was great stuff—old fashioned hard work meets new social marketing. It was all so exciting! My company seemed poised to become a small American success story. And the success wasn't stopping just with me and my company. Our vendors and suppliers—many of which are small, family-owned companies like ours—were growing with us, just the way we dreamed it would happen; just the way it should be in America, at least in the America of the past.*

As I have since learned, times have clearly changed. Working hard and doing things the "right way" doesn't really seem to work anymore, at least not from my experience. Just as we started to take off, manufacturers in China decided to come in and counterfeit our product, virtually bringing our international business to its knees overnight, causing me and my children sleepless nights of trying to remove the hundreds of counterfeit listings from Alibaba and AliExpress, and costing us our sales.

Being a fighter and knowing this was wrong, our small company, comprised mostly of family and a few close friends, started putting in lots of time and money—hiring attorneys, signing up for accounts to try to report the counterfeits on Amazon and eBay and Alibaba—basically doing anything we could to protect ourselves, but to no avail. For lack of a better phrase, it has been like playing the arcade game Whac-A-Mole; we whack one counterfeit listing down and 100 more counterfeit listings pop up. Some counterfeits were not very good but, to the naked eye, others were almost indistinguishable from our product. Some of the companies even duplicated my hand writing on the "Thank You" card we send with each purchase. In fact, they even included my personal information—Facebook page and telephone number—on the card, so that I now field dozens of phone calls, texts and Facebook messages each week from angry consumers who thought they had purchased our product, only to find out they purchased from a foreign company that has nothing to do with us.

Research revealed that we are not alone and that we have zero chance of stopping this. Big companies deal with the same thing and, of course, they have more money and resources to fight it. But why should they have to? As we talked with various people it was rather surprising to learn about other small companies that were forced to shut down because knock-offs manufactured in China basically put them out of business.

I have no issues with companies that choose to manufacture in China, or with goods imported from China that are made legally and do not infringe upon another person's rights. But counterfeits are a whole different situation, as they make it impossible for someone to capitalize on his or her own hard work and ingenuity. Indeed, this can only serve to deter people from even trying.

So what can we do? Who can we turn to for help? We have done everything legally and the only way we know how, the good old American way. What can we do as a small company to save ourselves and our product? Unfortunately, it seems very little. At the very least, we as Americans can educate ourselves, watch out for counterfeits and knock-offs that make their way into our market, and make sure not to buy them.

Linda Maria Gomez
CEO & Founder of Fullips, LLC

Linda built a business while the rest of the world was collapsing. She reinvented herself in her 50s. She came up with an idea for a product, patented the design, trademark and copyright protected her brand, built a prototype, figured out manufacturing. She worked hard to build her sales, and finally—after years of effort and expense—began to reap the rewards. Unfortunately, she had stepped into the IP cyclone at a time when working hard and doing things the right way no longer counted for much.

As wrong as all of this is, the problem goes beyond simple theft. For the safety of consumers, all legitimate *Fullips* products are made from food-grade plastic and FDA-approved colorant, manufactured and packaged by two reputable companies in the U.S.A. While food-grade plastic is assured to be free from dyes and recycled plastic that may be harmful to humans, the materials and manufacturing process used in the *Fullips* counterfeit copies are entirely unknown. While these counterfeits may be visually similar or even visually identical, right down to—as Linda said—the handwritten note in the product's packaging, the fact remains that they are made from unknown and potentiality dangerous materials. The fear that an unwitting customer might suffer an adverse health effect due to one of these counterfeits, Linda told me, plagues her constantly.

And yet, despite the clear-cut, black-and-white wrongness of the situation, little is being done on the macro level to confront the issue. As with a legitimate products, counterfeits thrive or fail based on their ability to readily connect with the end customer—thrive or fail based on their site visibility and distribution capability—and in this way the platform itself necessarily becomes culpable in the scheme. So what is a market giant like Amazon doing to combat the problem? In a word, not nearly enough. As stated, The Counterfeit Report has identified and reported tens of thousands of counterfeit products like Linda's to Amazon for listing removal, and yet many of these listings were not removed—even after repeated reports.[89] This, despite Amazon's claim that "the sale of counterfeit products, including any products that have been illegally replicated, reproduced, or manufactured, is strictly prohibited."[90]

What was that phrase ol' W. liked to use in the lead up to the war with Iraq? How did it go? That Saddam was "giving aid and comfort to the enemy?" If that's what it takes to invade Baghdad then I say we declare Amazon an IP WMD and we get some damn boots on the ground. If someone like Linda—someone who so exactly embodies the American spirit of self-determinism, hard work, and innovation—can't make it out of the mire we're sinking in, then surely our country—our "way of life"—is under threat.

You can find the ongoing report on Linda's product, including images of the picture-perfect counterfeits and resources for where to report a fake, at https://thecounterfeitreport.com/product/585/Lip-Enhancers.html.

INDIE INVENTOR JOSH MALONE

As is probably clear from the above, Linda's story really got to me. My own experiences with the CardShark had damn near brought me to my knees, and it outright staggered me to imagine going through something that was, on so many levels, even worse. Here was a single mom and breadwinner who, facing down the fear of losing everything in the 2008 crash, dug deep and invented a solid product, protected it from pillar to post, pushed it through the monu-

---

[89] "Amazon Counterfeit Practices Put Consumers at Risk," The Counterfeit Report, February 16, 2017. https://www.thecounterfeitreport.com/press_release_details.php/?date=2017-02-16&id=623

[90] "What is Amazon Associates Anti-Counterfeit Policy?" Amazon.com. https://affiliate-program.amazon.com/help/node/topic/GER4LUCFFTZJ2FDC

mental obstacles to market actualization and finally started to achieve success, only to have the rug pulled out from under her by counterfeiters who not only siphoned off the lion's share of her business but also degraded her company's reputation in the process. It made my blood boil to think that, despite the clear-cut, black-and-white wrongness of the whole situation—and despite the clear-cut steps that platform giants like Amazon could take to proactively combat the problem, *steps they themselves acknowledged the need to take*—little was actually being done.

As I was going to find out, though, given the unfortunate state of affairs, stories like Linda's were a dime a dozen. Case in point is indie number two: Josh Malone, an inventor whose patented idea wasn't so much counterfeited as it was usurped. I was introduced to Josh by Josh's U.S. Inventor compatriot Paul Morinville, who promised in his introduction that Josh's story was "a doozy." Paul had skillfully covered the particulars of Josh's story in a 2017 article he penned for IP Watchdog called "Water Balloons: Weapons of Mass Destruction in PTAB,"[91] and has generously given me permission to reprint it here:

Josh Malone has eight kids. On hot Texas days, he and his kids would enjoy a water balloon fight to cool things off. Josh is normally in the rear with the gear. He is the family reloader, filling and tying water balloons to supply his kids with the ammunition necessary to keep the backyard action going. It was during one of these skirmishes that Josh figured he could replace himself if he just created a weapon of mass destruction. He thought of several ways of doing it and then, like so many inventors before him, he obsessively tinkered until he finally invented one that worked. It screws onto a garden hose and has dozens of long tubes. Attached to the end of each tube is a self-sealing balloon. You just turn on the hose and when the balloons are substantially filled,

91 "Water Balloons: Weapons of Mass Destruction in PTAB," Paul Morinville, IP Watchdog, January 27, 2017. https://www.ipwatchdog.com/2017/01/27/water-balloons-weapons-mass-destruction-ptab/id=77637/

you shake them, they fall off, and the kids launch another attack. Leonardo da Vinci would be proud.

He named it "Bunch O Balloons." Josh knew then that he had a winner, and building a company based on his own invention became his American Dream. He filed a provisional patent application in February, 2014. Things went quickly at the U.S. Patent and Trademark Office (USPTO) and his first patent was issued about 18 months after the provisional application was filed.

Patenting proved Josh was the inventor and that he had an exclusive right to his invention. But more importantly, the patent could be collateralized to attract investment to build his startup. Investors look at upside potential and downside risks. On the upside, a patent's exclusive right meant that if Bunch O Balloons took off, Josh would be able to keep competition at bay long enough to establish his startup and return the investment. On the downside, a large company with deep pockets, existing customers, and solid distribution capabilities could steal the invention and massively commercialize it, thus flooding the market and killing Josh's startup. But patents mitigated this risk. In the worst case, Josh's investors could take control of the patent and return their investment by defending it against the same infringers who killed the company.

Josh manufactured an initial batch of products and then ran a crowd-funding campaign on Kickstarter. This campaign was a hit, generating 598 orders on day one and bringing in nearly a million dollars overall. Within a couple of days it triggered national media coverage in Sports Illustrated, Time, Good Morning America, *and the* Today Show. *Bunch O Balloons went viral, with 9.6 million YouTube views. I can only imagine how Josh must have felt… this would mean everything to his growing family.*

Over the next few months, orders kept pouring in. He was contacted by several ethical manufacturers seeking to license his invention. With business picking up fast, Josh partnered with a company called ZURU, who is now marketing, manufacturing and selling Bunch O Balloons. Josh achieved the American Dream. But that means nothing under the current American patent system.

Kickstarter is regularly watched by potential investors, customers, and ethical businesses. But there are others. Infringers also monitor Kickstarter for potential new products and as it turns out, the better a product does on Kickstarter, the more likely it will get knocked off. Bunch O Balloons got knocked off by TeleBrands just a few months after Josh launched his Kickstarter campaign...

Today, the U.S. patent system favors infringers like TeleBrands. In fact, it is a CEO's fiduciary duty to steal patented technologies, massively commercialize them, and then never talk to the inventor unless they sue. In the vast majority of cases inventors cannot access the courts because contingent fee attorneys and investors have largely left the business, so in most cases the infringer gets to keep the invention free of charge.

Much has been written about how Congress in the America Invents Act of 2011 stacked the deck against inventors by creating the Patent Trial and Appeal Board (PTAB) in the USPTO. The PTAB turned property rights upside down by immediately invalidating the property right already granted by the USPTO and then forcing the inventor to re-prove the validity of the same property right. Under the leadership of Michelle Lee,[92] the deck was stacked even further by setting PTAB evaluation standards much lower than the court. Michelle Lee's decision to set these low standards weaponized the PTAB for the mass destruction of patents. And a weapon of mass destruction they certainly are. The vast majority of patents evaluated in the PTAB are either invalidated or neutered. Big infringing corporations know this.

So when Josh sued TeleBrands for patent infringement, TeleBrands responded by filing a PTAB procedure called Post Grant Review (PGR).[93] The court did not stay the case pending the outcome of the PGR, and ordered a

---

[92] As I said in a prior footnote, we're going to talk much more about Michelle Lee in Part IV. For now, stay tuned.

[93] Kip's footnote: Sorry for the confusion. A PGR is functionally equivalent to an IPR: the PGR was created by the AIA, is a way for a third party to challenge the validity of an issued patent, and takes place before a panel of three APJs within the PTAB. The distinction is largely based on timing; according to LegalAdvantage.net, a "PGR can be filed immediately after patent issuance or reissuance, and IPR can only be filed after the period for post-grant review has passed or if no PGR is filed then nine months from the date the patent is issued (or reissued)." See https://www.legaladvantage.net/blog/what-is-a-post-grant-review-pgr-ipr-pgr/ for details. Had the period in question elapsed, Josh would have been IPR'd instead.

preliminary injunction against TeleBrands. TeleBrands appealed the prelimi-
nary injunction to the Federal Circuit.

During the pendency of the appeal, the PTAB rendered its verdict—Josh's
patent was invalidated as indefinite under Section 112. The claims state that
the balloon must be "substantially filled," which according to the PTAB is not
defined: "... the Specification does not supply an objective standard for mea-
suring the scope of the term 'filled' or 'substantially filled.'"

But how else can you write the claims? You could use grams of water if a
balloon was a solid structure, or perhaps if all balloons were exactly the
same. But manufacturing processes that make balloons are not accurate pro-
cesses. The thickness of the balloon's wall varies greatly from balloon to bal-
loon and even in the same balloon. Yet Michelle Lee's PTAB invalidated the
patent that Michelle Lee's USPTO had just issued. (Five other patents have
been issued to Josh, one even refers to this very PTAB proceeding as prior art,
yet it was still granted by the examiner. I kid you not.)

The Federal Circuit, while deciding a preliminary injunction was properly
granted, addressed the PTAB decision in its oral arguments and in its deci-
sion. In oral arguments Judge Moore stated, "You have to be able to say sub-
stantially, 'cause there's a million patents that use the word substantially."
And in their written decision the Federal Circuit explained: "We find it diffi-
cult to believe that a person with an associate's degree in a science or engi-
neering discipline who had read the specification and relevant prosecution
history would be unable to determine with reasonable certainty when a wa-
ter balloon is 'substantially filled.' Indeed. I suspect that all eight of Josh's kids
can do that too."

Josh's case is not over. Already he's spent multiples of what he earned in his
Kickstarter campaign and probably everything he's made in this entire Ameri-
can Dream. Yet, he's got years left of litigation and millions more to spend.

Patents can be invalidated in multiple ways by different branches of gov-
ernment and under different standards. Often these branches and standards
disagree with each other, as is the case here. Today nobody can know if a
patent is valid until the Federal Circuit or the Supreme Court says it is.

But this is the world inventors live in. If you invent something marketable, you will pay for it with years in court and millions of dollars. Nobody respects patent rights. They don't have to. It is better to steal them and litigate the inventor into oblivion. Josh is fortunate to have a partner willing to fight with him and accept considerable financial burden. But most inventors cannot even open the courthouse doors.

If that story makes your blood boil, you'll be glad to know that on November 21, 2017, a jury in the Eastern District of Texas returned a finding of willful infringement of Josh's patents by Telebrands, upheld the validity of the patents, and awarded $12.3 million in damages to ZURU Ltd.[94] Still, this outcome should only highlight the fact that, lacking the full weight of the legal resources ZURU put behind the effort, Josh would have been unable to substantively defend his patent at all. Lacking ZURU's backing, his defeat would have been nearly a foregone conclusion—a fact his infringers were very likely counting on. Rare as Josh's story of victory is, perhaps rarer still is a company like ZURU: one willing—even in a patent landscape badly mutated by one-sided legislation—to both invest and fight for that investment. As I've described, investors like ZURU and the litigators like those they employed to fight for Josh's patents are fleeing the space in droves, and one has to wonder if anyone will be left to step up when the next Bunch O Balloons comes around… or if, even as you're reading this, the last one has already fled the space.

Nowadays Josh Malone has taken his experience as an aggressively infringed inventor and become a diehard advocate for the small inventors. Along with his work at U.S. Inventor, Josh has dedicated his time and energy to walking the halls of the D.C. circuit of politicians and lawmakers to try to be the "hope and

---

[94] "Telebrands loses $12.3 million verdict for willful patent infringement of Bunch O Balloons" Steve Brachmann, IP Watchdog, December 8, 2017. https://www.ipwatchdog.com/2017/12/08/telebrands-loses-willful-patent-infringement-bunch-o-balloons/id=90797/

change,"[95] and we indie inventors are all indebted to Josh for his tireless efforts. You can find out more about Josh, his story, and the important work he is doing—and get involved yourself—at https://usinventor.org/josh-malone/.

INDIE INVENTOR ADAM ULLMAN

My work as a screenwriter is driven by one theme: the underdog in the underbelly. Nothing cuts me to the core like the story of someone thrown up against impossible and unjust odds in an upside-down world where such an injustice is par for the course: where "right" and "wrong," "fair" and "unfair," no longer count for much. Given that, once you hear it, I think you'll understand why Adam Ullman's story made this list.

Adam's story was brought to my attention by a fellow-in-arms against the inequities of the patent system over at US Inventor. The introduction went like this: "Kip, meet Adam Ullman. Adam is a third-generation inventor who took one of his dad's inventions and commercialized it. He was knocked off by the Chinese, who are now importing their knockoffs into the U.S. The cost of litigating against these infringers is very high, so Adam is pushed into a bad position having to choose between pursuing the infringers and continuing to operate his company. Adam has spent tens of thousands of dollars so far to protect his patented property, and the situation is far from settled."

Adam's invention is the SteriShoe®, an ultraviolet light-emitting insert that you put in your skanky sneakers to kill that left-in-the-bottom-of-your-hockey-bag-in-a-gym-locker smell. It's also good if you have toenail fungus, recurrent athlete's foot, or diabetes. Not only did Adam have a patented idea, a trade-

---

[95] Hey Obama, good line, but you should have been paying more attention to the inventors out there slogging to do just that rather than the Google and Big tech machines that just deep-throated you with all sorts of *change*.

marked company name, packaging, a distinct look, and a website, but he also was manufacturing his product and selling it exclusively under his exclusive patented rights... and then things went wrong. I'll let Adam tell you all about it himself in the pages that follow. In addition to being an inventor and an entrepreneur, Adam also has a Master of Laws in Intellectual Property, Commerce, and Technology, so his take on his own experience—and the current state of the patent system in general—is a particularly well-considered and well-informed one, and one which he was gracious enough to share as well.

Here's Adam:

I'm a third-generation inventor. My maternal grandfather had 17 patents, my father has his name on seven, and I have my name on a bunch. After bringing other products to market I decided to go to law school and study business law and intellectual property law. While I was in law school I was reviewing my father's 6th patent, which described a shoe tree containing a UV light source to clean the inside of shoes. I was taking a look at that, and thinking, "OK, that makes sense," and after law school I decided to build a business around it. Citing his patent as prior art I successfully got 20 global patents on UV shoe sanitizers—utility patents in the US and abroad, and a couple design patents as well.

We launched the product first to the podiatric community. The feeling was that a doctor's recommendation is the highest recommendation a product can receive: we launch to those professionals and get the buy-in from them and then that would give us success going into retail.

The product started off really well, and then somewhere in our supply chain in China somebody decided to start using counterfeit components. I found this out because all of a sudden we were having a high failure rate in the field. I had to figure this out. I had product that was already in the States, more on the water, and we were building another production run. I went over to China to catch that next production run and was assured that everything was fine. Still, something wasn't right. I reached out to TDK, the Japanese elec-

tronics company whose capacitors we were allegedly buying—the capacitors in the products needed to be a certain material—and I took a picture of the capacitors in the supply run in China and I sent that to them, and my phone immediately rang. They said, "These are absolutely not our parts." They asked me to send them some samples. I did that. They did a microscopic cross-sectional analysis of these components and determined that what were supposed to be their $0.30 capacitors—which are expensive for capacitors—were counterfeit. So somewhere in the supply chain somebody saved a couple dimes here and there.

A few dimes might not sound like much, but the impact from this switch was huge. It should have put us out of business, actually. The product failure rate in the field was over 25%. My shareholders stepped up with a loan to the business, but it cost me hundreds of thousands of dollars, stunted growth, and lost us a lot of the goodwill and reputation we had with the brand. And the company being tied up in that situation opened the door for patent-infringing products to sell on Amazon. And Amazon... I've got a love/hate relationship with them. Amazon is a great channel for revenue, but at the same time it's also a very easy way for infringing companies to sell without repercussions for their patent infringement. I don't know if you are aware of this or not, but Amazon—under our patent laws—is not a seller. They are a marketplace. So one of the things that typically incentivizes or disincentivizes retailers from selling patent infringing products is, once they are on notice, they are liable for what's called "treble damages," which can be 3x damages. So they don't want to get involved in a patent dispute. Now Amazon has no such concern, because they are not a seller. So these patent-infringing products sell on Amazon and Amazon collects its commission for every sale that happens on Amazon. That's how Amazon works. It's a big issue in our laws.

Frustrated by everything that was going on, especially with my knowledge of IP, I started going to patent conferences in the Bay Area. I was pretty vocal. I put on a seminar at the Licensing Executive Society's Annual Conference one year, and at one of my patent licensing conferences I was speaking with somebody from Amazon—I happened to be in the lunch line next to him—and

I told him, "Look, I love you guys and hate you." He said, "What's up?" And I explained the situation to him and he gave me his card and said to reach out. And through that contact I had some conversations with some attorneys at Amazon, and Amazon now has something called a "neutral infringement patent evaluation program." And through that I've had some success in having some infringing products de-listed. But it's a process.

I'm a firm believer that innovation is key to the American dream. But the thing about that is, innovation is really hard. Copying is really easy. And what's happened here is that our borders have opened up. We talk about IP theft internationally, like, "We can't have that happen there," but we allow it to run rampant on our shores. That is a big issue, especially for a small business. If you take a look at Article I, Section 8, Clause 8 of the Constitution, that's the clause that delegates Congress's authority to set up the Patent and Trademark Office. "Securing for limited times to Authors and Inventors the exclusive Right to their respective Writings and Discoveries" Now if you look at Article 1, Section 8, you see that the other clauses listed enable Congress to declare war, coin currency, prosecute counterfeiters, set up post offices and roads. It's major, major powers enumerated through Article 1, Section 8. So the fact that patents are mentioned in there gives them significant weight, and it gives you a sense of what, really, Congress should be doing to protect our intellectual property rights. If a foreign nation is copying my patent, there's an argument to be made that Congress should react to that as strongly as if that nation were counterfeiting our currency. It's all delegated under that same provision. But what's happened here is that, not only do we not have that, but litigation and the things you need to do to actually defend your patent yourself are all incredibly expensive. And as a small business owner, why should I have to go through such an expense to defend a right that I was awarded?

My education and my belief point to our patent system being a bargained-for exchange with society. Rather than something being a trade secret, where if the inventor of a cure for cancer gets killed by a bus it's gone forever, the patent system is set up wherein, by disclosing to society the invention, you have the exclusive right to practice that invention for a period of time. Now I will also say this: patents are generally pretty easy to work around. They are usually very specific in nature. So what happens is, if that person documents a cure for cancer and publishes it, other scientists can now look at it and say, "Hey, wait, couldn't we do this a different way? Couldn't we do this here instead?" And that spurs more innovation. And that is actually the driving factor for our patent system. It increases creativity.

If you take a look at your different areas of intellectual property, you see that patents have a term limit. They expire 21 years after date of filing.[96] It used to be 17 years from the date of issue, but that changed because people were doing what's called a "submarine patent." They would keep the prosecution of the patent open for literally decades. Then, finally, when the patent issues, they go back to other entities operating in the space and say, "You're infringing, you're infringing." And that created big issues. So it's 21 years from the date of filing. Now if Congress wanted to, they could shorten the patent term. That seems totally reasonable. Or maybe in certain spaces you leave it the same length, but in other spaces you shorten it. In software you could shorten it. In biotech, where you have to go through FDA regulations and things like that, clinical trials that take years and years, maybe 21 years is the right amount of time. But Congress can control that per the Constitution. Exclusive rights are secured for a "limited period of time." So what is that time? You can segregate different industries and classifications accordingly. Copyrights are not in perpetuity. The term of a copyright does keep getting longer and longer—I think the last count I had was something like 105 years. It's been a while since I looked at that, but that's roughly what I remember from law school. Trademarks are valid as long as they're used in commerce. If

---

[96] For clarity: "As a general rule, utility patents filed after June 8, 1995, have a life of 20 years from the date of filing of the earliest non-provisional application to which it claims priority. It is important to note that the term of a utility patent is tied to the earliest filing date of a non-provisional patent application to which it claims priority. The consequence is that by filing a provisional patent application as the priority application, then filing the non-provisional patent application one year later, you are able to file a patent application whose lifespan will be 21 years from the earliest filing date." From "How Long Does a Patent Last?" Richard's Patent Law. https://www.richardspatentlaw.com/faq/how-long-does-a-patent-last/

you've got a trademark and you're actively using it and monitoring its use, that trademark is protected. The point being that we do have different carve-outs for different things. But patents are intentionally designed to expire after a certain period of time. And again: that is for the benefit of society. So somebody has the right to commercialize and not face direct competition for their patent. It's a right to exclude, and we should make this easier. The incentive is paid off in the benefit on the other end. So it shouldn't be such a challenge. It shouldn't be such an incredible challenge to protect what you have disclosed for the betterment of society.

It's a frustrating situation. With my SteriShoe product, we started out actually doing some clinical studies. I had a peer-reviewed article published in the "Journal of the American Podiatric Medical Association" that then other patent infringing products started to cite. I did the heavy lifting and got this done at an expense, and then others were using it. And their data wasn't the same! Our trials were of our product with our components. Theirs would have been totally different. But with the way the situation is, it's very hard to stop things like that. Also, if we want to take a look at China: I don't know if you've ever ordered anything from eBay or Alibaba or anything like that—it might take a while for it to ship to you, but have you ever noticed what those shipping rates are? They're incredibly low, and that's because the Chinese government will subsidize the shipping rates. If I want to ship something to China, it's going to cost me a fair amount of money. But if a Chinese company wants to send something to the U.S., it's a couple dollars. So when we talk about IP theft—or when our politicians talk about IP theft—the conversation is focused on what happens outside of this country. We're not looking at the actual practical ways in which the IP theft happens here. China is subsidizing the shipping so that infringing products can come in very inexpensively, and the process to stop something at our border is very, very difficult. So it's things like that that make it really challenging to innovate. They make it really challenging for the U.S. inventor. Look at what's happening with Kickstarter. People are launching Kickstarter campaigns, and let's say they're running a 60-day campaign. Well 30 days later there's a factory in China already selling the product. They've seen the product and they've seen the demand and they can just start shipping it out.

Again: innovation is hard, copying is easy. Think about branding. Branding, which is where you develop your trademark, requires attention to detail, a control of the quality of your goods, and marketing dollars. How do we justify marketing dollars other than a sale? Those dollars help build the brand. And so you start to see how trademark value is critical. It's a slightly different perspective on IP. IP is the intangibles on a balance sheet, or in the marketplace. Why do companies trade at an earnings multiplier? It's the intangibles. It's based on what we're expecting them to do based on their IP and their brainpower. That is the reason why a company trades at a multiplier of its earnings. Anything above a standard return on investment expectation is based on the intangibles. It's brain power and intellectual property. Think about how much of our markets are comprised of intellectual property value, if you look at it from that standpoint. How much of the market is based on value based on what a company is working on, what their brand represents, and what their brain power will do in the future. It's trillions of dollars.

What's happened now is that the America Invents Act actually made a patent a potential liability, rather than an asset. It used to be set up where, if you had somebody who was infringing your patent, you could put them on notice. At which point, after you'd put them on notice, they could go to court for a declaratory judgement. They could say, "OK, we want to know if we are infringing this or not." If they don't do anything then the patent's still there and it's kind of assumed that they know that they're infringing. Until you actually made a claim against them, though, there was nothing they could do. They couldn't go after your patents. One of the things that the AIA did was allow for a company to just randomly challenge the validity of a patent. So if, let's say, Google wanted to try to invalidate one of my patents, they could just file with the PTAB and IPR my patent. And imagine that at this point I'm just trying to get my business up and running, keep it moving, and now I need to spend probably $50,000 to $500,000 to defend the validity of a patent that already went through proper prosecution with the US government for it to be *granted! This turns something that should be an asset into a liability. So for me, the inventor, where is my incentive to actually file a patent? Where is my incentive to share my innovation for the betterment of society? It's now kind of disappeared and in many cases may be better protected as a trade secret.*

In terms of this entire situation with patent trolls: yes, there are bad actors out there. There are bad actors in everything in society. The problem is that if I'm a software developer, and I get a patent, and all of a sudden somebody says, "Hey, I'm going to IPR your patent, I don't think this is a valid patent," and I don't have the money to defend it, I'm going to need to sell it. I'm going to need to try to sell it as a hopeful asset rather than a liability. So I might sell it to an organization that knows nothing about software development, but they've got the deep pockets to defend that patent. Now Google or Apple or another Big Tech company says, "Wait a minute, you're a patent troll. You just bought up a patent and you're not practicing it." Kind of a double-edged sword there, isn't it? Because what created what? So I think we need to change the way we look at defending a patent which has been granted by the U.S. government and has gone though proper patent prosecution.

The situation right now is a mess. You think you've got an asset and all of a sudden it's turned into a liability. What just happened? Much less being on your own to defend it. If there is somebody counterfeiting our currency, that's our government's problem. They have to defend that. I made a bargained-for exchange with the government that granted my patent, so why does this pre-sumable asset become a liability? And this is where things have shifted in a very negative way, and why we are actually losing our edge in innovation to foreign nations.

And unfortunately our politicians, in this case, I think miss the mark and don't understand—at least from a small business perspective or an individual inventor perspective—the challenges you face when bringing an innovation to market. Somebody could have something on the back burner, could be working on it a couple hours a week, and it may take them years to actually get it to market. Sometimes we come up with things and the technology needs to catch up. Cutting edge technology is very expensive, but five years down the road it's much more affordable. Somebody might be working a day job and saying, "Look, I'm going to put away a little bit of money every year, and when I can I'm going to go and do this." And they know that the clock is ticking on their patent. It doesn't make them a patent troll, if they go after somebody who infringes their patent. But what's happened here is that the Goliaths, these corporations, have shifted patent law because they don't want to be restricted by a patent from an individual software developer or a small company. They lobbied heavily and they got reform, but it was done in a very short-sighted way. The smaller individuals and the smaller businesses were not considered.

One time I called the Patent and Trademark Office and I had a representative tell me that patent litigation is "the sport of kings." That's actually a statement from someone at the USPTO's patent helpline: "It's the sport of kings." Shocking, right? This from the Office that is supposed to guarantee the very protections that their representative is saying is reserved for the kings who can play the game. And again, Amazon, makes it really easy to deal with trademark infringement, with copyright infringement. Those things are black and white. Does it look like the same thing? Yes. Design patent, still black and white. Utility patents are a gray area. There is interpretation. Now Amazon has stepped up in terms of that "neutral patent evaluation program," which is great. It's kind of what I said to the Amazon representatives we need. We need a kind of intermediary to say, "OK, is this infringing or not?" A way to get the infringing items taken down. So it is a decent system. Prior to that, I had better luck getting things taken down from Alibaba than Amazon. Kind of crazy.

I am a firm believer in the intellectual property system. It's something I care about and would love to see us get corrected. I am not doing the daily fight anymore, though. I really was for a while, trying to push for legislative reform—not as much as some, but what I could do with a wife and two kids and trying to make a living. But whatever I can do to help, I want to help. I want to help make people aware of the challenges to small businesses and individual inventors. I want to help make them aware of why the patent system is important.

In the end I had to step down from the daily activities of my shoe sanitizer business. I couldn't rely on that being my income. That business needs the proceeds from every sale to stabilize and take on the challenges it's still facing. The business is still going, but I've moved on to other ventures. It's frustrating. It's not what I expected.

Adam is now the Principal at PDCIP, LLC, his consulting company where he provides interim executive leadership advising businesses on a variety of issues, and is the Interim CEO of The SiTa Foundation, a "non-profit organization that is developing a small, discreet, cellular-based technology device that prepares victims of violence to rise up against repeated abuse by empowering them, building confidence, and increasing safety through the use of technology that engages allies." If you'd like information on how to support SiTa's efforts, you can contact Adam through LinkedIn.

And if you're interested in the (legitimate) SteriShoe shoe sanitizer—and let's face it, you should be—you can find it online at www.SteriShoe.com

INDIE INVENTOR M. DAVID HOYLE
(B.E. TECHNOLOGIES)

You may recall me mentioning M. David Hoyle at the beginning of Part II: may recall me telling you that technologies invented by David and protected by patents issued to his Tennessee-based company, B.E. Technologies, were stolen by Google. Now I'm going to tell you how. If you guessed that it has something to do with an IPR, you're catching on... but it's worse than you think.

And let me reiterate, if I haven't already made it clear: if anybody embodies the David-versus-Goliath nature of the fight we indie inventors are in, that somebody is M. David Hoyle.

I first met David through some mutual acquaintances in the scrum: other indie inventors neck-deep in the fight against the seemingly endless stream of injustices handed down by today's USPTO. These contacts knew that I was writing a book about what was going on, and they pointed to David as someone who could help me sort through and effectively convey the convoluted story of our dying patent system. After all: what David went through could go down as *the* emblematic illustration of what is so very, very wrong with the system today.

See, you probably know Google as the friendly search engine that helps you find the closest yoga studio or the best banana bread recipe. You put in what you want, it kicks back the best result, and all's well with the world, right? Not quite. Despite the focus on—and the mired mysteries of—its search capabilities and its results algorithm (ask anyone working in SEO and they'll tell you just how *proprietary* proprietary can be), the fact is that the core of Google's business is its *advertising* revenue. In fact, of the company's $22.5 billion in quarterly revenue in 2017, $19.8 billion was generated from advertising revenue alone—nearly 90% of Google's entire quarterly revenue.[97] As Gene Quinn and Steve Brachmann state in their 2017 IP Watchdog article "PTAB Invalidates Targeted Advertising Patents, Preserving Billions in Google Ad Revenue," "Google's ability to target audiences with its advertising tools... is of crucial importance to the entire company—It's just too bad that Google and Alphabet do not own the patent covering the technology which is earning them about $20 billion each quarter."[98]

In September 2012, David's B.E. Technologies, LLC, filed a lawsuit alleging infringement by Google of two of the company's patents. The patents in question both pertained to targeted advertising and both listed David as the inven-

---

[97] United States Securities and Exchange Commission Form 10-Q for Alphabet Inc. for the quarterly period ended September 30, 2017. https://abc.xyz/investor/static/pdf/20171026_alphabet_10Q.pdf

[98] "PTAB Invalidates Targeted Advertising Patents, Preserving Billions in Google Ad Revenue," Gene Quinn & Steve Brachmann, IP Watchdog, November 13, 2017. https://www.ipwatchdog.com/2017/11/13/ptab-targeted-advertising-patents-google-ad-revenue/id=90076/

tor: U.S. Patent No. 6,628,314, titled "Computer Interface Method and Apparatus with Targeted Advertising," issued to B.E. Technologies in September 2003, covering the system of upgrading software for an application providing advertising over the internet, and U.S. Patent No. 6,771,290, titled "Computer Interface Method and Apparatus with Portable Network Organization System and Targeted Advertising," issued in August 2004 to B.E. Technologies and covering a toolbar application developed by B.E. Technologies which provides a user with a graphical user interface (GUI) to interact with the data system providing an advertising platform. With valid patents in hand, David and B.E. Technologies had been actively target-advertising its audiences since long before Google knew what that even meant. And yet what was the outcome of B.E.'s infringement lawsuit against Google? You know the story already. A retaliatory IPR quickly followed and, in December 2013, the PTAB handed down their final decision: 100% of all challenged claims were un-patentable. David's lawsuit was dead in the water: his patents had been led down to the river bank and double-tapped by the PTAB's hired guns. Today, David's patents are gone and his company is shuttered, while Google continues to enjoy the fruits of the empire it built on his back.

To paraphrase David: "Robbers break into my home and steal everything. I'm reeling because I've got nothing left. At least I know who did it. But then, as I'm getting ready to bring action against them, the robbers turn around and sue me for having owned the house in the first place. We go to court, and the judge rules in *their* favor, and awards them my house."

"Google has been successful at stealing vast quantities of others' intellectual properties without paying one cent," he told me. "At the same time, they have successfully convinced the media, the general public, Washington, and the Courts that they are the true innovators, and are victimized by nefarious forces who, in their words, 'don't make or sell any real product or service.' But the so-called culprits that Google is calling out are hardworking individuals trying to start businesses and create new technologies. Many have mortgaged all they own to [do this]. While these true innovators struggle to survive and be heard and overcome Google's media echo chamber, Google enjoys handsome profits from the inventors' stolen technologies, allowing them to afford the best lawyers in the country. In contrast, most inventors cannot afford proper representation, no matter how blatant Google has been in the theft."

Bad as this is, it takes on an entirely new aspect when viewed in the context of the AIA's history... and the not-so secret fling between Obama's White House and Google that produced it. In fact, given the billions in revenues that were at stake, one wonders whether the looming prospect of David's lawsuit wasn't foremost on Johanna Shelton's mind during her 128 visits to the Obama White House during the crafting of the legislation; indeed, given the percentage of Google's overall revenues that these technologies represented (and still represent), one wonders whether she was thinking about anything else.

Recent discoveries and revelations have cast this entire era in a new light, and pending litigation brought by David against current and former employees of the USPTO intends to prove a level of complicity between Google and the USPTO previously suspected but unsubstantiated, all of which we will discuss in more detail in Part IV. For now, let David and B.E. Technologies stand as another victim of the PTAB: another patent holder abandoned by the Office whose protection it had been guaranteed.

INDIE INVENTOR GENE LUOMA

Please meet Gene Luoma, an indie inventor from Duluth, Minnesota. I first heard about Gene from a fellow inventor whose voice kind of cracked when he mentioned Gene. You'll see why. Gene invented the Zip-It Drain Snake,[99] a

---

[99] www.zipitclean.com

simple but brilliant tool for clearing stopped-up drains. I have my own experiences with the Zip-It: with my long hair and my three long-maned daughters, Gene's invention has been an absolute lifesaver on many a Saturday night in our house. There have been times when the prepping and primping and vying for the shower and the mirror over the sink has bordered on an extreme sport, and before the Zip-It I had Roto-Rooter on speed dial. As far as I'm concerned, Gene is a genius and a saint.

Gene told the story of his invention—and what happened to it—in an article first published on IP Watchdog.[100] With his kind permission, that article is reprinted below:

The Zip-It drain cleaning tool was an invention that I developed around the year 2000 because I was getting tired of my bathroom drains getting clogged. My daughter Kim didn't like haircuts and when she would take showers, her long hair would clog the drain and cause the water to back up. I tried to solve the problem by using liquid drain chemicals, plungers and bent coat hangers, all to no avail. There were no products available on the market that provided a quick remedy for clogged drains.

One morning, after dealing with yet another slow-running drain, I went into my garage and found a worn-out plastic sled hanging on the wall. I took it down and my idea was to cut a 24-inch-long strip from it which was narrow enough to fit down the drain, and then I fashioned a handle on one end and cut barbs along each side of the long strip. I took it into the house, stuck it down the shower drain, pulled it out and the amount of hair it pulled from the drain looked like a dead rat had come out of there. The Zip-It was born!

The Zip-It is a simple, one-piece, injection-molded product which is composed as a flexible strip having barbs and then a handle on one end which has an aperture both for finger gripping and for hanging. I searched for prod-

---

[100] "A Personal Plea From the Zip-It Inventor to Support the Inventor Protection Act," Gene Luoma, IP Watchdog, September 2, 2018. https://www.ipwatchdog.com/2018/09/02/personal-plea-zip-inventor-support-inventor-protection-act/id=100804/

ucts similar [to] this and didn't find any. My next step was to file for a design patent and a utility patent to cover this device. The product has no assembly: when the Zip-It drops out of the injection-molding machine, it is ready to use.

I took my prototype to a large, big box home supply company and showed it to their plumbing buyer. He was amazed and said that his store would purchase thousands of units if they were available in such large numbers. I licensed my product to Cobra Products, a division of BrassCraft Manufacturing, which is owned by MASCO Corporation. 10 years and 12 million Zip-Its later, Cobra's plumbing sales representative, Doug Cohen, was let go from the company and went to work as the president of G.T. Water. It was at that time that G.T. Water began selling a product that infringed upon my patent for the Zip-It.

Cobra Products filed a lawsuit for patent infringement. G.T. Water then filed for a re-examination of my patent at the U.S. Patent and Trademark Office to invalidate my patent claims. My licensing agreement with Cobra and BrassCraft was to share equally in the cost of defending my patent. However, Cobra Products elected not to help me in the defense of my patent at the USPTO. I was forced to bear the total cost of that alone. After seven years of validity proceedings, which cost more than $250,000, the Patent Trial and Appeal Board (PTAB) invalidated all 12 of my claims covering the Zip-It. How can the USPTO issue a patent with 12 claims and then use the PTAB to neuter my patent? Since this has been going back and forth with the court system, there are now numerous other entities which have engaged in copying and infringing my patented invention.

Defending the patent is only part of this particular battle. I was born with muscular dystrophy and two of my children suffer from this same disability. My family suffers from a form of muscular dystrophy called Facioscapulohumeral Muscular Dystrophy, which is a terrible, progressive disease that requires adaptive equipment to keep us independent, mobile and able to stay in our homes as the disease worsens each year. The royalties I had been earning from licensing my patent have been helping everyone in my family stay independent despite our disabilities. This income had prevented me from becom-

ing reliant on the government for assistance. Because of the invalidation of my patent claims, however, the chances of me being able to continue making an income from this invention are gone. With the loss of this important income, it has threatened the future independence for me and my family.

Gene's story represents a dark—and far more common—counterpoint to Josh Malone's. Whereas Josh's licensor funded the efforts to defend his Bunch O Balloons patent, Gene's licensor did not. Whereas Josh had the backing of lawyers paid for by that licensor, Gene was left to find and—because the contingency firms, knowing the likely outcome of an IPR proceeding, steered clear—fund his own legal representation. The system created by what M. David Hoyle has called the "worldwide scheme of the most powerful multinational corporations... to monopolize technologies by destroying patents and innovation"[101] did what it was designed to do: it hung the little guy out to dry.

I called Gene the other day to check in on my friend and comrade-in-arms in the fight for patent rights. He told me how, in its heyday, Zip-It Clean was selling on average of 300,000 units a month to major big box stores, and bringing him an average of $40,000 a month in royalties—money that was paying the medical bills for Gene and his children. After Gene's patent was invalidated, he told me, the original licensor, Cobra—who, it should be noted, breached their contract by refusing to help litigate the patent—refused to pay him another cent in royalties. Instead, insult to injury, they filed a lawsuit against him, trying to claw back every penny they had ever paid him. To date, Gene has spent over $400,000 to fight back, but to no avail. Indie inventor and indie inventor advocate Josh Malone even tried to help, hiring a troop of lawyers to work on Gene's case, but the efforts so far have been fruitless. The natural term of Gene's Zip-It patent expired in 2021, and Gene estimates that, between the time of his last royalty payment in 2018 and the patent's expiration, he lost

---

[101] "Is Unified Patents a War Profiteer?" M. David Hoyle, IP Watchdog, March 31, 2020. https://www.ipwatchdog.com/2020/03/31/unified-patents-war-profiteer/id=120267/

over a million in royalties from what he estimates to have been several million units sold.

Gene, like his son, is wheelchair-bound by his illness. Gene is 78 years old.

—

In the next few pages I'm going to introduce you to some of the brilliant minds operating in the IP space today. I'm grateful to each and every one of them: they all took time out of their busy schedules to try to help me understand the IP world as it stands now, what to tell you to make sure you know what's going on, and what it's going to take to get us back on track.

IP EXPERT MATTEO SABATTINI

First up is Matteo Sabattini. Matteo was one of the first sources I found and started following when I embarked on this project. Matteo, in addition to holding a PhD, an MBA, and (at the time of my first meeting with him) the position of CTO at Sisvel Corp., is a dedicated and conscientious contributor to the ongoing debate about how to build a healthy innovation ecosystem. I sought him out after I read his article "NPEs vs Patent Trolls: How to Build a Healthy Innovation Ecosystem,"[102] on IP Watchdog. Matteo's deep dive into the issue had me pacing around the room: he was describing something that cut right to the heart of my CardShark predicament.

---

[102] "NPEs vs Patent Trolls: How to Build a Healthy Innovation Ecosystem," Matteo Sabattini, IP Watchdog, February 4, 2015. https://www.ipwatchdog.com/2015/02/04/npe-patent-trolls-innovation-ecosystem/id=54427/

To understand why this distinction between NPEs and patent trolls matters, we have to remember that the AIA was pushed through largely on Big Tech's claim that patent trolls—those IP-owning third-party entities who "bully the market by asserting, or threatening to assert in court invalid or bogus patent portfolios to industry players that do not have the resources to defend themselves or for which it does not make economic sense to fight back in court"[103]— were both a threat to and a real-time drag on the engine of American innovation. The changes that the AIA enacted—specifically the PTAB and its various post-grant review processes, most notably the IPR—were designed to be explicitly antagonistic toward such entities and their payday-seeking litigation. But of course, not all third-party IP-holders engage in this practice. Nor is all third-party IP litigation abusive: indeed, patent licensing and monetization firms provide an essential service to corporations, research centers, inventors, and the whole innovation ecosystem by, among other things, pursuing legitimate claims against infringers, providing a revenue stream for the inventor to reinvest into their own company's further innovation. Such NPEs serve a vital and value-adding function in—to borrow Matteo's term—a "healthy innovation ecosystem." But what happens when a legitimate patent claim from a legitimate representative is lumped in with the chaff?

This was part of the situation I was facing with the CardShark. Like a lot of inventors, I was leveraged to the hilt in just gaining and then defending my patents: the spare capital required to actually compete for market share on my own patented idea—to develop, produce, distribute, promote, et cetera, et cetera, my own invention—was nowhere to be found. I'd been forced by circumstance to become an NPE, defending my patent rights but not actually producing anything myself, and accordingly had found myself more than once on the wrong side of that convenient lack of distinction between NPE and troll.[104]

---

[103] *Ibid.*

[104] The various names I've been called by infringers and infringers' legal council could fill a second book. Until that book is written, believe me when I tell you that nearly *all* of those exchanges contained some hint or accusation that, in their view, I was just a troll doing what trolls do. I've included an account of the most frustrating and/or egregious example in the next section. Again, for now, stay tuned.

But, as Matteo says in his article, patent trolls "extort, in aggregate, significant sums" by "creating risk and exploiting the exorbitant costs of litigation (especially in the U.S.)"[105]; this was certainly not what I was doing. In fact, if anything, the realities of my situation were more exactly represented by a second phenomenon Matteo describes in his article, that of the "Patent Ogre." The Patent Ogre, as outlined, is "a large company that has a significant market position in a product or service category and protects its economic interest by suppressing, bullying and/or simply grinding into the ground smaller, more innovative competitors that have patented technologies. Faced with a small innovator with patents that potentially read on its products or services, the patent ogre... may refuse to license the technology at market rates... create publicity campaigns to label the inventors as trolls, and drag them through endless legal maneuvers until they run out of money... Then the patent ogre continues to derive economic benefit from the technology that someone else invented or perfected."[106] From where I was sitting, that sounded an awful lot like my experience with the various infringers I'd already run up against myself: big ogres throwing their weight around and taking what they wanted with no regard for the people whose lives they were upsetting. I felt like Matteo "got it" and I gathered up my nerve and reached out, hoping for an interview but not actually expecting to hear back. I was thrilled when he very kindly obliged.

Matteo grew up in Bologna, Italy, so after figuring out the timing and establishing that I *parlo italiano*, we dove right into important stuff... like the Ferrari factory in Maranello and the Ducati factory just up the road over there in Bologna. I think the fact that I had actually done a few laps around the Ferrari test track in Maranello (I told you I covered racing, right?) might

105 "NPEs vs Patent Trolls: How to Build a Healthy Innovation Ecosystem," Matteo Sabattini, IP Watchdog, February 4, 2015. https://www.ipwatchdog.com/2015/02/04/npe-patent-trolls-innovation-ecosystem/id=54427/

106 Though this quotation comes from the same article, Matteo is himself quoting from an article published on IAM-Media.com by Michael Gulliford. "Who Is Uglier? Patent Trolls or Patent Ogres?" Michael Gulliford, IAM-Media.com, December 16, 2014. http://www.iam-magazine.com/blog/Detail.aspx?g=dba41734-acfa-4759-93fc-26fcb2c0b98a

have surprised (impressed?) him. Maybe not. Either way, Matteo was more than gracious with his time and his substantial knowledge in the IP space. The most pertinent parts of our conversation are included below.[107]

Kip: *Tell me: What do you call the little guy who spent all the money getting the patents and can't afford to compete with the big guys in the big box stores, and therefore can't manufacture the invention?*

Matteo: *I would call that a really good question, for starters.*

Kip: *It just feels like the independent inventors are getting hammered by the patent system even more hardcore for the past five years.*

Matteo: *Yaaasss.[108] So true. Lately, because of the changes in legislation, what we're seeing is a lot of independent inventors [trying and failing] to stay in the game against the big corporations. It's [creating a lot] of "I would if I could" inventors.*

Kip: *In your ["NPEs vs Patent Trolls"] article you refer to "patent ogres": these large entities who bully IP holders and the market. It seems like, when you talk about independent inventors trying to "stay in the game" against the big corporations, it's this behavior you're talking about. Like this is the real issue that needs to be addressed on a legislative level.*

Matteo: *Yaasss. The sort of overly broad legislation that we have now—legis-*

---

[107] It should be noted that the views, opinions and statements by the interviewee are his alone, and do not represent the views of his current or former employers, nor any employee thereof. Furthermore, the views expressed herein were current at the time of the interview and may not represent the views and opinions of the interviewee at the time of publishing, in light of the ever-evolving jurisprudence and policy/legislative landscape.

[108] Italians love their long *"yaasss."* So dramatic.

lation that fails to distinguish between legitimate, reasonable, and appropriate enforcement activities on the one hand and the illegal and inappropriate extortion-like activities of trolls on the other hand—bludgeons the entire technology industry. It's a blunt instrument that's trying to address a situation that's very complex. It's a simple answer to a complicated problem, and because of that it's the wrong answer a lot of the time.

Let's just make one point very clear: the burden to defend the patents against IPR weighs very heavily in favor of the large corporations who can bring to bear this process—[a process] that can take upwards of 18 months and which, while under review, often blocks the inventor from being able to go forth and enforce [their patent] against any other infringers. It also tends to side with the company bringing the IPR against the patent holder.

Kip: *Yaaasss...*

My conversation with Matteo highlighted once again the harsh reality that today, under the AIA, an indie inventor's patent is only as strong as the money that indie has to defend it, and that, even with these resources at their disposal, the indie's path is far from certain. There is simply nothing (save the expense, nominal to a major industry player) to stop a larger company or a consortium of larger companies from hitting our intrepid indie with an IPR, at best hamstringing them for the course of the 18-month proceeding and at worst invalidating their patent—or its most significant claims—altogether.

Lamenting the current state of affairs, Matteo left me with this gem: The indie inventor is "creative in ways that might see things ahead of their time, ahead of the curve. They may create the curve. [They may be] 'a genius for [whom] time has not yet come,' as Michelangelo said. [They] risk all for that which hasn't been invented yet, but to also face the outright theft [happening now] is going to end [the endeavoring of] even the courageous few who live to forge new ideas into hard-won realities."

IP EXPERT WILL PLUT

Clicking my way through the various online forums, I came across a podcast hash-tagged "patents matter" and an episode featuring our next IP expert, William Plut. What grabbed my attention was Will's passionate—I'd even say borderline visceral—response to some of the antics going on in the patent world: according to Will, there were a lot of clandestine "shenanigans" going on, and he was straight up not having it.

The case in point in the podcast interview involved a phony claim and a phony brochure to back it up: Company A, casting about for a means to cut Patent Holder B (and whatever licensing fees Company A owed Company B) from their bottom line, had photoshopped an entire brochure of prior art to bolster a false claim that their concept pre-dated Company B's patent filing date. Let me tell you: this set Will off. This was some serious clandestine bull-shit—some "shenanigans" of the highest order—and Will had plenty to say on the subject, and on the current state of IP affairs in general. I hoped that, if I asked real nice, he'd say some of it to me. I clicked over to Twitter, found his profile, and reached out.[109]

Striking out on Twitter (did you read the footnote?), I swizzled over to LinkedIn and hit pay dirt: Will not only replied to but also readily accepted my interview ask. I prepped for our scheduled chat with reams of paper and a stack of pens, ready to go old-school journalist on him… then I picked up the phone and got hit with a firehose. If the interviews I'd conducted before were,

---

[109] Truth be told, and since we're being all "full disclosure" here, my first attempt didn't go so well: I replied to one of Will's tweets with a request to interview him for this book, and somewhere in the tubes and sprockets of the internet that tweet took a strange turn, and I was hit back by a gaggle of hash-tagging HotEuroBitches who wanted me to join them for a threesome. Maybe it goes without saying that I was a complete novice in the land of Twitter: for longer than I'd care to admit I didn't even know that "#" was a "hashtag" and not a "numbers sign," so I kept referring to things like #patentsmatter as "number sign patents matter"… Quoth my kids: "Ma, yer hopeless." #JesusHChristonatricycle

say, equivalent to some light recreational drug use (you're at a party, some-one's got something, you only live once...), then this was mainlining meth. I'd hit the motherlode.

Quick background on Will, before we jump in, unapologetically cobbled together from a few sources online:

Will has been involved in invention monetization and the patent space since 1997. He received a masters in Engineering in Robotics and moved to Silicon Valley (Silicon Valley girls are, like, totally smarter?) in 1998. He prac-ticed prosecution for over eight years at a leading Silicon Valley prosecution firm (he's also a member of the U.S. Patent Bar) and then went on to found Patent Profit International LLC, an organization specializing in the sale of top-tier patents. He is the lead inventor on over 35 patents and patent applications, has successfully monetized his own inventions, and has founded several Silicon Valley startups. His inventions cover everything from robotics to laser-based projectors (he sold that) to power conservation for smartphones (he sold that technology—technology which is in every Samsung smartphone today—to Sam-sung). Will recently launched Verasuit LLC, a company specializing in advanced medical and infectious disease protective gear.[110]

...all of which implies but none of which states explicitly the fact that what Will really is is one of the 30 or so people worldwide who knows all the secret trap doors and hidden passageways in the patent labyrinth: from the USPTO head officers to the lobbyists to the big companies who out-n-out infringe to the shell corporations set up by those same big companies to front as patent trolls on those companies' behalf so they can go out and squash the big com-panies' competition... Whatever it is, Will has seen it and is calling bullshit.

Lunchtime wasn't going to deter him, either. He dropped some serious knowledge on me between hoagie bites. To include it all would run the rest of this book; I've included the pertinent highlights. Dig it:

---

110 This guy: *suuuuch* an underachiever, amIright?

Kip: *Can you explain your Twitter handle: "Arms Dealer in the Patent World"?*

Will: *To understand what I do, you kind of have to understand the situation with how IP litigation works in general. I'm sure at this point you know all about patent trolls and what they do, right? So you understand basically how a patent can be weaponized. Without getting too far into the weeds, we can generally say that big companies do the same thing against each other. There's this constant race and back-and-forth for market share and new tech and new software and everything else, and part of the way those battles are fought is through IP litigation. So when you have patent wars going full force between these F500 companies—when you have these big companies going after each other with IP infringement litigation—I'm the guy who stands in the middle and sells litigation-grade patents to whichever side needs them.*

Kip: *OK, so how is what you're describing—where companies buy up patents and then go after each other with IP litigation—different from or related to patent trolling? Because it sounds a lot like patent trolling.*

Will: *Absolutely. And that's the thing that people need to really understand about the whole patent troll thing. These are not two different things, these are one and the same thing. The patent wars have sort of tapered off now, but during the real heavy part of it—for maybe five years before and after the AIA —the vast, vast majority of the money being spent buying up patents was coming from the Big Tech companies, but very few of those patents actually got assigned to those big companies. I'll repeat that. Big Tech bought the assets, but the majority of those assets went into shell companies in Nevada or in some other jurisdiction where the Big Tech company could obscure ownership. So the question you ask is, Why would they do this? And one big reason is that it allowed them to avoid attribution. If Microsoft put patents into a little shell company and then that shell company sued a company that Microsoft didn't like, or maybe even a startup that Microsoft saw as a potential threat to their market share, the reporters would scream, "Patent trolls are evil! Look at* *what they're doing! They're killing startups!" By assigning the assets, a company like Microsoft could skirt the issue. The shell company would take all the heat with no blowback on Microsoft. There was no bad press for the mothership. Which, since the late '90s, had been a big problem. Back in the late '90s, IBM was suing companies and getting all of this bad press, and the other Big Tech companies saw that and they said, "You know what? We can get around this pitfall if we just hide the ball a little bit." And often I was the guy in the middle. Some startup would go belly up and the Silicon Valley venture capitalists would punt the assets—they'd come to me and they'd say, "OK, Will, just get a couple hundred thousand or a million or two million dollars for the IP." They already took a $20 million write-off on the loss, so the deal was just to get something for the startup's founder. Selling the assets wasn't going to cover the write-off. So I would go and sell the assets, usually to the big companies. The big companies have all the money, right? And in over 100 deals, less than five actually went to the mothership. Less than five sets of assets actually went to the mothership company itself. When you assign the asset a lot of less sophisticated companies will assign it right into the big company. That's the exception, not the norm. 95 times out of 100 a big company puts the assets in a shell company. I remember one time negotiating this sale with Google, and six months later I'm looking around and I see that asset at the center of a lawsuit that a shell company has brought against seven other companies. And there's the media watching this and saying, "Patent trolls are evil!" And I'm like, "But that's not a patent troll! That's Google!" So that's the first thing it did. It allowed these Big Tech companies to hide and avoid attribution. It kept these Big Tech companies from being blamed and getting the bad press. But the second thing it did was it dovetailed into this wonderful lobbying effort that these Big Tech companies were building in Washington. Because at the same time, these Big Tech companies really were tired of dealing with these different entities coming to them and threatening IP litigation and demanding licenses. And there was this narrative that was already in the public consciousness about how patent trolls were killing startups. There were all of these sob stories around Silicon Valley about these great startups that had* *been killed in the crib by trolls. Big Tech figured, "Wow, we can use this." And they came into Washington with all of this data saying that 700 startups had been sued that year and 300 went right out of business because they couldn't afford the litigation. Because understand: a patent lawsuit is $1 to $5 million. If you're a startup on $500,000 of funding, you're dead in the water. So there was a lot of data behind the patent troll narrative. There were just a lot of writers already going on this. So these companies started building this lobbying effort around this narrative, trying to change the patent laws. Which, you have to understand, was just another in a series of runs they'd taken at Congress. They'd been trying to change the laws all along. In '07 they'd been close but they failed when they tried to change the Continuation laws. They got kicked back. So the next time they came around in Congress they came in with all this data saying, "Hey, 700 startups got sued this year and 300 went right out of business because they couldn't afford the lawsuit." And they got the change they wanted, which was the AIA in 2011. Which, of course, included the PTAB. And the PTAB, in one sentence, effectively killed small patent portfolios. It decimated the value of a single asset. Because if you're a big company and I come knocking on your door saying, "Hey, you've infringed my whatever," and I ask for anything more than $200,000, why would you pay me? You can haul me in front of the PTAB and just destroy the asset for $200,000.*

But here's the thing about all of this, and here's the point that's the answer to your question: literally 80%-90% of the so-called "patent troll" litigation was litigation brought by shell companies backed by Big Tech. In 2014 I actually built what I call the assignment database. After enough deals you see patterns. So when Microsoft, for example, bought a patent portfolio and stuck it into a shell, I could look at the shell and go, "Yeah, that's Microsoft." You do these deals often enough and you're able to detect all the signals. And sometimes they're not obvious—sometimes you need five or six to weave them together and eventually go, "Yeah, that one's Intel, that one's Apple." But I could do it. There are maybe 10 to 12 ways to see through a shell. One way is to track the assignment language and detect which of the assignments are en-

demic to each large corporation. It's like with what you do. Let's use you as the example. You're a writer. Say there's a specific story—one single and specific story—but there are many different ways to tell that story. You're a writer, you have your version of that story. It's the same thing with lawyers. Each lawyer has his or her own spin on how to write a patent assignment. And you can be sure that same lawyer loves the smell of their own work, and generally does not change their assignment over time. Each time you had your lawyer for the CardShark file a new assignment for the patent office, I guarantee he didn't change his assignment language. This is the same thing that happens at a Big Tech company like Microsoft. They have their own unique words in their assignment, as does Google, and Apple, et cetera. You get the picture. So when the same assignment shows up in some random small Texas corporation with a strip mall address, it's a safe bet that you're dealing with a shell corporation fronting for the large company that uses that same assignment language. So I took this information and I built this assignment database to kind of see through all the shell companies. And I automated this thing. I hired a computer programmer kid for a summer one year and he just built this beautiful code. We literally identified all of the assignments for every big company and then every shell and matched them. And then we looked around at all of this so-called "patent troll" litigation, and 80% of it was coming from Big Tech. It was coming from companies either fronting for or backed by Big Tech companies. And right around this time I had a conversation with a company called Allied Security Trust, or AST. AST is like a consortium of 30 big companies—Intel, Amazon, Google, Apple: all these guys all working together to defend against patent owners. AST had the only other similar database at the time called "Patent Freedom." And I said, "According to my numbers, 80% of these troll lawsuits are backed by Big Tech companies. Either the Big Tech company gave them the assets or the Big Tech company gave them the money for the litigation. And we compared notes, and their numbers were very similar. It became apparent that the whole patent troll feudal war thing has just been Big Companies slapping each other all along. The Big Tech companies kind of went, "OK, we got what we wanted lobbying in D.C., we changed the patent laws, let's start dropping the whole patent troll thing." And it kind of started withering away around 2016. Today it's a fraction of what it was in its heyday.

Kip: *Good God. So it wasn't just that the patent troll narrative was pushed by Big Tech, it was actually created by Big Tech?*

Will: *Exactly. They essentially created the problem and then they sold a story about that problem so that they could sell a solution for that problem. Let me give you another example. The biggest patent troll of all time is a firm by the name of Intellectual Ventures. IV was founded by a guy by the name of Peter Dekin who, ironically, was actually the guy who coined the term "patent troll" in the first place. He coined the term while he was working as a patent attorney for Intel. There was some shell company that was suing Intel at the time, and "patent troll" was what Peter started calling them. Peter went on and founded IV with Nathan Myhrvold, one of the Microsoft greats, and they started with more than a billion dollars. $200 million came from Microsoft, $200 million from Apple, $200 million from Intel, $100 million from eBay, $100 million from Google. Myhrvold pitched in money. And as you can probably imagine, that kind of money buys a lot of patents. And IV brought countless lawsuits. And that's just what went to court. You have to figure that the vast majority of licensing got worked out without court, so God only knows how many people they went after in total. And IV was, in part, Microsoft's attack dog. The Valley is flowing with stories of startups who were just mauled by IV at Microsoft's behest. And then after all of this IV litigation, who turned around and complained to Congress about the horrible ills of patent trolls? Microsoft, Apple, Intel, and Google. You get all of these people testifying in front of Congress about the evils of patent trolls, talking about how patent trolls are the reason that the laws have to be changed, and I'm sitting there saying, "Wait a second, you guys created Frankenstein's monster!" They created the monster, sent the monster to maim competitors and startups, and then they held up pictures of the Mummy and Dracula and they said, "See? We need* *to change the patent laws!" Because remember: the PTAB was pitched by these big companies as a cure to the ill of patent trolls like IV. So again: they basically created a problem and a story line about that problem so that they could sell a solution that was a Trojan horse for a weapon they could use against any upstart competitor. Because realize, too: if you have 100 patents and $100 million, the PTAB does nothing to you. If you have one patent and $100,000, you're dead.*

In my view, the whole thing was basically Big Tech's retread of the 1970s Big Tobacco play. It's the big money pitching a very skewed version of the truth to the policymakers. And this includes things like paying for biased research, just like Big Tobacco did. If you go back to the Big Tobacco play of the '70s: they had fake researchers at universities that they were paying, they had writers on the payroll pushing articles attacking regulation and reporting on how safe smoking was. With Big Tech you've got the same thing. You've got secretly-paid-for researchers producing bogus studies. And there's layers within that. On one side you've got willing accomplices intentionally producing biased materials, and on another you've got these well-meaning but unwitting suckers being guided and incentivized and otherwise influenced to produce biased materials based on the biased information they've been fed. Probably the most effective of the willing accomplices was Mark Lemley. Mark was a Stanford professor and a paid attorney for Google who literally said, "I am not an unbiased observer" when it came to the patent situation, but whose studies on the so-called "evils of patent trolls" were used in Congress in its consideration of various patent reform bills, including the AIA. You have a guy who's firm is on Google's payroll and who is also producing these studies for Google's benefit via his Stanford professorship, and then Google turns around and presents it like some smart and impartial academic just happens to agree with them on what sort of reform is needed. They go in front of Congress and they say, "See? This smart, impartial academic agrees with what we're saying." And Mark is just the tip of the iceberg. So that's on one side. On the other side you have these unwitting suckers who Google just sort of steers. The poster-child for unwitting suckers is probably Colleen Chien. Colleen was

*a legal professor from Santa Clara University, and Colleen has published nu-
merous anti-patent troll papers and testified before Congress on the subject,
all of which would be fine, except for the fact that Santa Clara's law school
received Google money for legal research. Academic ethics dictates that one
should disclose funding from a source when the source is a subject of the research, but Colleen's papers neglected to mention Google funding, while
Google as a defendant was certainly a subject of any patent troll research.
Again, though, she's a different animal from someone like Lemley. A Google
executive once described Colleen to me as "a nice person; nice and naive."*

Kip: *Talk to me about the money that Big Tech puts into Washington. Because
it really seems like the AIA wasn't just a case of Big Tech making a skewed
argument well. It seems like it was sort of a case of Big Tech pitching to policymakers who'd already been sort of induced to be open to the message.*

Will: *Yeah, and that's a whole other can of worms. Big Tech sold the story, but
who bought it and why did they buy it? Maybe there are things amiss in the
patent system, but all of that is nothing compared to the things that are amiss
in the political system in general. It's nothing compared to the political machinery that makes it all possible.*

*I'll give you a story that'll maybe help illustrate the situation for your
readers. I was talking to the head of government relations for one of the Big
Tech companies one time, back in maybe 2013 or 2014, and I asked him, "So
how much money are you guys putting into Washington these days?" And he
said, "Eh, less than you think." I said, "Give me a number. Ballpark." And he
hemmed and hawed and then he said, "Well, officially, $20 million. Unoffi-
cially, $100 million." And I was sort of stunned. And then he said, "Look, Will,
it's simple. You ever play roulette? 50% of the time you land on red, 50% of the
time you land on black. If you could bet on both with 100x return on each,
then you'd never lose. You always come out way, way ahead. It's the same on
the Hill. You give 50% of your political campaign funding to the blue and 50%
to the red, and whichever side becomes the loss doesn't matter. Both return*

*100x." So you have to think about it like that. 50 million to each side is going
to get some things accomplished up on the Hill and, while that sounds like a
lot, you have to realize that large patent licenses used to cost billions each.
Legislation that kills patents, and thus the need for those licenses, saved these
Big Tech companies billions in the end.*

Kip: *So where are we now? What is your sense of the current state of the
patent system?*

Will: *The biggest thing is, it's too easy to destroy a patent. The patent system
was supposed to be a quid pro quo: an inventor trades publication of his invention for an exclusive right. That quid pro quo is basically gone. Injunctions are near extinct, and the cost to invalidate a patent is too low. Look at
it from the perspective of a big company: if a patent owner asks for a license
that costs more than $200,000, then why would the big company ever pay that
license? An IPR costs less than $200,000. The PTAB has become rational con-
sumer behavior for the big company in a patent system that gives the option
to destroy a patent for $200,000.*

*Again, the thing that's so important to realize in all of this is that the PTAB
is only effective against small patent portfolios. If someone were to try to use
the PTAB against 50 or 100 patents, it wouldn't work. Not all of the assets
would be entered into proceedings, it would be too hard to get prior art on
that many patents, it would be too expensive to put that many into a leaky
strategy en masse, and so on. The PTAB did not undermine the ability of big
companies to wield their large patent portfolios. So it's not the big companies
with the big portfolios who've been impacted. The patent system has been
shredded, true, but the impact is felt by the little guy almost exclusively. You
need 50 patents and $50 million to play now. The AIA basically turned the
patent game into what the British had 200 years ago: a thing only for the rich.
For 200 years the American system was democratized. The little guy could get
a patent, the little guy could get funding. Think about Nikola Tesla. Tesla never sold a product in his life. All he did was sell and license IP. And he's still to*

this day probably one of the best engineers of all time. But without the patent system, Tesla's just a frustrated employee. He never gets around the corner on anything. But what happens to the next Tesla, in the era of the AIA and the PTAB? You can't fundraise on a patent, now. So what do you do? You go back to some 9-to-5 job. You say, "Screw it."

The thing is, though, at this point, you can't even really blame the big companies for doing what they do. The thing you realize when you're in Silicon Valley for a long time, and you're playing with the top tier, is that it's all just business. You step back and you become very impartial and you say, "Yeah, from a business perspective what they're doing is actually the smart play." It's shrewd business. If you're a big multinational who can do this, you almost have an argument that you've got an obligation to do it. It's just kind of how big business is, now. If you give them the opportunity to do it, they will do it. If you can just destroy some competitor's asset for $200,000, you're not doing your job for your bosses or your shareholders or whoever if you buy it or license it for $500,000. So you could argue that, in a system that permits the PTAB, it's rational consumer behavior for a big company to act the way they do right now. It's rational behavior in a system that has evolved to permit that behavior. The problem isn't the practice, it's the laws that dictate the rules of the practice. The problem is the laws being changed in the first place. Once you set down that path, you can't blame the big companies for doing this. If you don't like how these companies are playing the game, you have to change the rules.

It does get really dark when you think about it, though. It gets really dark when you think about the fact how we got here. When you think about the fact that a big company would sue a startup with a shell, avoid attribution, everybody would scream, "Patent troll!" and then that big company used that narrative to change the laws, but then at the same time that startup that got destroyed would come to me and I would be charged with selling the assets, and then the same big company would buy the assets for 10 cents on the dollar! It's so nefarious. But again: it's just business. It's big boys using what muscle and what leverage they have. They have the money, they have the re-

sources, they have 40-50 patent attorneys... It's just big boys doing what they do. If you're Genghis Khan's army and there's some little village that has something you want, you steamroll the thing. That's what you do.

Kip: *So what do we do?*

Will: *Among other things, we start by reinvesting in the patent office. One of the things that gets missed in all of the noise around the patent troll narrative is the fact that there really are bad patents being issued, and bad patents that have been issued that should never have been issued in the first place. And that's a problem for everybody. It's a problem for the patent holder and it's a problem for anybody that patent holder wants to bring litigation against. Because think of it like this. In the upstairs bathroom in your house, there's a leaky pipe. You hire a plumber to fix it. It costs you $600. You pay the plumber and the plumber leaves. Months later you come to learn that the pipe wasn't fixed at all, and all that water was leaking down into the kitchen below, the walls, the whole house. The plumber's going to charge you $25,000 to come back and fix it all. The secondary damage is much more money. You have to pay much more to fix the leak that you thought you already had fixed. This is the IPR process. The IPR is largely there to clean up after the USPTO's poor and rushed examination of your patent. They do a hatchet job, you get approval, then BOOM, all these years later, you see how much more it's going to cost you. And this happens in part because there's very little accountability in the realm of the patent office, with the ones who did the hatchet job in the first place. So part of the solution is to fix the examination process: to give examiners more time per case and fewer cases to review. Let them get it right in the first place so everyone downstream walks on more certain footing. Then, of course, destroy the PTAB.*

So good for you, Kip, for writing this book. Books like this are what we need. We need more things that really try to expose what's going on and try to democratize the message of the patent system and how to protect IP for the entrepreneurs. I really think that's why TV shows like Shark Tank *are so popu-*

lar, and especially among the teens. Because it's the American Dream. Stories that depict relative rags to riches. And their access point is TV, which is a platform that fully democratizes the knowledge base for all to see and experience and understand. And that's important, because it is everybody's right. It's everybody's right to be able to have an idea and protect that idea and work hard and make their dreams come true. People need to know that, and they need to understand that that's what these big corporations and their lobbyists have been working and still are working to take away.

Follow Will Plut, Arms Dealer in the Patent World and PPE supplier through the COVID-19 pandemic, on Twitter at https://twitter.com/WillPlut.

IP EXPERT ERIC HURWITZ

At some point in my own patent odyssey, six degrees of something or other brought me in contact with Eric Hurwitz of Mensch Capital and The License Bureau ("a patent and brand licensing agency with a diverse set of clients including NASA Astronaut Buzz Aldrin, American Express, Warner Music Group, and Miss America"[111]). Eric has been in my corner ever since, helping me squeeze blood from stones, i.e. convert infringers to licensees. I don't know why (or honestly, how) he's stuck with me all these years. He's been on calls with me where the response I got from an infringing company brought down the power grid behind my eyes. Does he have some morbid fascination with watching me drown ever-so-slowly in this patent quicksand? I guess it doesn't matter why, so long as he's there to fish me back out again...

All of which is to say: Thank you, Sweet Baby Jesus, for Eric.

Given our history and his background, Eric was definitely on my short list

---

[111] https://westfourip.com/team/

of interviewees, and he was good enough to sit down and share his views on the subject from the standpoint of his substantial knowledge and experience.

Kip: *First off: thanks, Eric, for your insights and your guidance, and thanks for doing this interview.*

Eric: *My pleasure.*

Kip: *I would like to make this book about solutions. Kvetching, but kvetching that points out the ways and the reasons that our patent system is broken. So tell us: Where are we?*

Eric: *Right now here's the deal. Small portfolios with one or two or three patents are not being bought up. Big money is buying up the big portfolios because these smaller portfolios are just too vulnerable and at high risk of being invalidated. The small portfolios are just too hard to monetize today.*

Kip: *Fun.*

Eric: *The IPR game has become a way of doing business. Things changed, spiraled for the smaller patent portfolios about the time you got in the game, 2010. In fact, your CardShark experience is a pretty good case study in how shitty a road it can be.*

Kip: *Hence the book.*

Eric: *Right on. Write on, right on. So let's talk about that. You've dealt with a lot of things, with the CardShark. One of the things you're dealing with—one of the strategies that's out there—is what's called "efficient infringement," which means that it is less expensive for large companies to use other peoples' or other companies' patented technology than it is to negotiate a fair license* *agreement and pay royalties. The bottom line is that today it is cheaper to infringe than it is to honor a license agreement and pay it out of the licensed product's proceeds. Once upon a time, there used to be an honor system. A company would sit down and negotiate. This is all gone. It stopped happening somewhere around 2013.*

Kip: *Why?*

Eric: *Economics. Large companies did the analysis and realized it was better to steal than do the right thing. Further, they were very successful in lobbying the government to weaken the patent system for their own benefit, using the narrative of the "Patent Troll" as the root of all evil in the patent ecosystem that needed to be dealt with no matter what the damage to small inventors, the U.S. innovation economy, et cetera. It has been very effective for the interests of large infringers and very bad for everyone else. Which leads to the subject of the IPR.*

Kip: *IPR is to small inventor as sugar is to diabetic?*

Eric: *I think, in practice, it has been bad for everyone who wishes to utilize their patents, whether they are practicing them (i.e. using them in the production and sale of their own products) or licensing them. But, yes, it has been particularly bad for small companies and inventors where it is very possible that in an IPR, which has insane invalidation rates, the inventor loses their entire patent portfolio in the blink of an eye. It just makes everything incredibly unfair. An inventor, such as yourself, does everything right, goes through the entire examination process at the USPTO, which as you know, is a difficult, time consuming, and expensive process, and then when that inventor goes to protect their government-issued property it gets invalidated in an administrative "trial"? It is such B.S.!*

Kip: *And what makes a bad patent?*

Eric: *That is a difficult question to answer because there isn't a single answer. There are a number of things that can and should legitimately invalidate a patent at trial, but in general I would say that bad patents are those that cannot be enforced because the claims are too broad or poorly written. That doesn't mean that those patents are necessarily invalid, it just means they're bad. One needs to be able to understand the invention based on the patent claims. Think of it like a plot of land. You can't own property if you can't delineate the boundaries. The boundaries determine the scope of the thing you own. Without that, you don't really own anything definite, which means you don't own anything. It's the same with patents. You have to know what your patent encompasses.*

Kip: *Another person I interviewed, Will Plut, said something similar. He talked about how the whole burden is brought to bear on the independent inventor, while there are no repercussions for the patent examiner. It feels like there should at least be a refund issued for all the fees to the inventor who was granted their patent only to have it IPR'd and revoked. If the patent wasn't valid then it shouldn't have been issued in the first place, and if it is valid then it shouldn't have been IPR'd!*

Eric: *Absolutely correct. There are no negative consequences for the USPTO and/or the patent examiner that allowed the patent. There is no security for the smaller patent holder, on either end. But what I find even harder to tolerate is that, when a corporation brings an IPR against a patent holder, a single patent judge decides whether that IPR petition is approved—then that same judge sits on a panel of three to review the validity of the patent in question. So automatically, going in, there's this bias. This supposedly objective ruling is coming from the three patent judges, one of whom just signed off on the validity of the IPR petition! It just seems completely unjust to me. It's no wonder an insanely high percentage of IPRs are ruled against the patent holder.*

Thanks again, Eric. I feel lucky to follow in your footsteps as you tiptoe us through the IPR minefield that is today's patent system.

IP EXPERTS LIZ STAPP AND TIM WOLF

Somewhere in the midst of all of this I received an invitation from Liz Stapp, Professor of Accounting & Business Law at the University of Colorado Boulder's Leeds Business School, to attend her class on business ethics and share my story with her students. I came by the invitation honestly: one of my daughters was attending the school at the time,[112] and had brought the CardShark's epic saga to her prof's attention. We would be joined by Tim Wolf, former Global CFO of Molson Coors and Chief Integration Officer of Miller Coors, who was there to share his views on business ethics based on his professional experience. I flew out and tagged along with daughter number three to a large, fluorescent-light-filled lecture hall crowded with all sorts of eager young business majors. The demographic was no surprise: in recent years Boulder had emerged as a hotspot for startup incubators, placing it, the school, and Liz firmly in the thick of the action.

Despite my status as a bona fide indie inventor, I was clearly out of my element: I thought the "EBITDA"[113] outburst from the kid behind me was his failed attempt at a self-contained sneeze. I flinched and checked the back of my hair, then offered him a "God bless you." Shows you what I know.

Luckily, no one seemed all that concerned with what I knew or didn't know

---

[112] Three of my kids, in fact, attended CU Boulder; that poor zip code the year they overlapped...

[113] "Earnings before interest, taxes, depreciation, and amortization (used as an indicator of the overall profitability of a business)." https://www.lexico.com/en/definition/EBITDA

about earnings before interest. They were there to hear from Liz, and once she started preaching all eyes were on her. For over an hour she absolutely threw down on the muddy history of capitalism and governance. Somewhere during the lecture she introduced me as a female inventor and asked me to share my CardShark story, and to relay some of the challenges facing indie inventors. It seemed a perfect starter course for the kids to understand in real time the never-ending game of Whac-A-Mole awaiting the indie inventor when, at one point, some of the kids actually held up their phones and showed me their credit-card-carrying cases: by now I'd gotten good at recognizing the different versions, and I could tell them that, sadly, these were mostly made by infringers and not CardShark licensees.

I'm pretty sure it was Liz's intention to "educate the shit" out of her class by presenting them with different business perspectives—that of the large corporation and the small inventor—but I think out of everyone there I got the most out of that day, given that I had the chance to sit down and huddle up with Liz and Tim in the empty lecture hall after the class filed out. We discussed the class, my book, and what needs to happen to keep the whole American IP thing from running off the rails as we continue barreling into the future. Among other things, Tim offered this absolutely priceless advice when I asked him, "Given everything we discussed in the class and everything you know now, what would you tell someone who's just starting out, hauling the water up the hill?

Tim: *I'd tell them that it is absolutely essential that you focus on the vetting process, in everything from how to build your team around you to how you eventually build your board and committees so that they drive home the single mission of your company or product. For the sole entrepreneur, this is a really hard slog. You have to really look outward, network, and connect to people both like-minded in your industry or sector, as well as branch out. For example: in building an executive board, there are some red flags you might want to avoid when heading out to fill the positions. Your board has to add* *value: who they know and the deals they can help make happen for you. No overboard executives. This refers to executives who are overboard-ed, meaning they are overcommitted to too many boards and cannot provide your company the attention, focus, and dedication to what you need, especially at the development and startup phase. Confucius says: the 10,000-mile journey starts with the first step. For startup inventors, this means that first step had better be well planted and firmly grounded.*

This advice hit me hard, and I wished in that moment that someone had said as much to me before I set off into the IP jungle with my CardShark and my faith and enthusiasm as my only real assets. In particular, I thought of my first lawyer: by now his inefficient, self-serving, and expensive way of operating had earned him the termination that, I have to suspect, he knew was coming. He seemed to take it pretty well, and rolled right into casual conversation about my future plans. I hung up feeling like the whole thing was settled… and then the other shoe dropped. All of a sudden the deals he'd cut for the CardShark—poorly and veeeeerrrrryyyy ssllllloooowwwwllly, I might add, hence the firing—weren't big enough to offset his $800-an-hour legal fees. He wanted more than what he signed for, and we were going to have to pay him to make him go away. We bit the bullet and drained our bank account, only to turn around the next minute and find ourselves sued for breach of an interminable contract that this lawyer had drafted *for himself.* I guess I should have seen it coming: lawyers are good at contracts, and when drafting one for themselves they might just slip in some cute 'lil clause or codicil that says that "if you terminate me with or without cause" I get to throw the whole thing—like a very expensive hand grenade—back on your head.[114]

---

[114] This infuriates me to this day. I was hiring a lawyer: was I supposed to know that I needed to hire a second lawyer to review the contract I was signing with the first lawyer? The whole thing feels predatory and oh-so-wrong, and it makes me sick to think that this asshole—or someone just like him—is out there pulling this same shit right now.

I said as much to Liz and Tim. Liz gave a knowing nod. "That sort of thing is why I do what I do," she told me. "I want to teach solutions to future entrepreneurs and business-generators."

With that in mind, she said, it was absolutely critical that I showed my book's readers that there were solutions. The situation, while challenging, was not unworkable. She pointed out that, while it was certainly true that patent laws had been more advantageous to indie inventors in the past, today's hyper-open information age made it possible for—for example—an indie inventor who is being crushed by an infringer to go on social media and effectively spread a serious firsthand account of the injustice, and rally the support needed to stay in the fight. Never before had the individual in general or the indie inventor in particular had such access to an audience as on today's web. She also pointed out that, unlike in the past, today's indie inventors had access to think tanks and entrepreneurial safe havens: places where like-minded creators and inventors could join their efforts with those of lawyers and engineers who would help them execute legal and technical documents or drawings for a percentage of the return on the work being done. Teams like this were (and are) much more cost-effective than the veritable solo missions that prior generations of indies generally had to embark on—a major consideration for any cash-strapped indie who, facing the mountain of fees and expenses associated with getting a product protected and produced, might feel inclined to throw up their hands and join the 9-to-5ers.

"Patents really are the last frontier," Liz told me. "Every other factor of production has been usurped. Lumber, oil, natural gas, farming... It's the one place left. And it's being assaulted on every side. The Founding Fathers could have never imagined the kind of collusion going on between big businesses and government against the lone inventors and innovators today. Never. The Founding Fathers did not want this. They didn't want class systems of wealth, and this, sadly, is exactly what we have. There is less and less to compete for, and the sad fact is that nowadays, unless you have the connections and the resources to influence the game, it's almost too tough. This is why so many startups fail." Still, she said, there is hope: "Incubators," she said, "are a very good solution for the inventors and the startup entrepreneurs."

Leaving our meeting, I clung to Liz's optimism. All of the interviews I'd done up to this point had only served to confirm my worst fears: that the patent world was fucked, perhaps irrevocably, and that everyone in a position to actually address the problems and the threats those problems presented were all drunk on Big Tech's KoolAid. It hadn't occurred to me that the very forces mutating the patent system—the rise of the internet and its associated tech monoliths—might actually hold the key to the solution: that the democratization of messaging, the ability to connect and collaborate with ease—to pool resources and abilities—might not only save the limping American innovation sphere from ruin but foster in a new and even more promising era... I wanted to believe it, but—given everything I'd heard and experienced—I still wasn't sure that I could.

PART IV

FOX IN THE HENHOUSE

"Market incumbents, typically large technology multinational corporations, could not be happier with the disintegration of U.S. patent law in the last decade since it enables them to free ride and to preserve extraordinary monopoly profits of typically $10 billion to $40 billion a year."
— *Neal Solomon, "The Disintegration of the American Patent System."*

"Ain't nobody here but us chickens."
—Louis Jordan and his Tympany Five, "Ain't Nobody Here but Us Chickens"

I. PEN PALS

By now, two things had become achingly clear to me. First: the situation desperately needed to change. Stories like Linda's and Gene's were maddening and heartbreaking on their own, but the idea that these stories were just a drop in the bucket and that, with no check on the existing legislation in place, that bucket was being added to daily by more drops and more buckets (and more buckets, and more buckets...) made my stomach turn. There was simply no other way to say it: The indie inventor was at war, waging a noble but woefully lopsided battle against a heavily armed adversary. So who were the enemy? The second point was now as clear to me as the first: for most of us indie inventors, the battle consisted of ducking, covering, and cowering from bombs lobbed at us from far away on Capitol Hill. Any one individual might survive to a greater or lesser degree, but strategically this method of engagement offered no possibility for victory, or even for improvement of our circumstances. For many, it was just delaying the inevitable. If we didn't engage with the source—if we didn't go to where they lived—then nothing was going to change.

But where was that? It was easy to talk in abstractions—Big Tech, the USP-TO, the PTAB, even specific companies like Google and Microsoft and Apple—but who were the players on the field? Decisions that have affected and continue to affect the lives of hundreds of thousands of independent patent holders in the U.S. had been made and were being made by individuals pushed and lobbied by other individuals. So who were those individuals? And what—if anything—could be done to make them see the damage being done by these decisions?

I honestly have no idea what they're teaching kids in school these days, but back when I was growing up our teachers told us that, if we didn't like something about the way our country was being run, we should write a letter to our political representatives and let them know our thoughts. This, we were taught, was our privilege and our responsibility as citizens of a country run by a government of, for, and by the People; it was also, we were told, an effective way to push for change. Lacking any clearer sense for where to start, I did just that: I sat down and I wrote to Kirsten Gillibrand, Democratic Senator from my home state of New York. I gave her an abbreviated version of my CardShark story, explaining how I'd done everything right—had filed and paid for the patents to protect my invention—only to run up against a system skewed to favor big infringers. I was hoping to convey the plight of the indie inventor struggling to stay alive while swimming in the shark-infested waters of the post-AIA patent system; also, as one of the mere 8% of independent patent holders who are women, I thought it might generally do some good to give a voice to the femme inventors. I dropped my letter in the mail, vaguely heartened by the thought that my story would be read by someone in a position to tackle the problems I and so many others were facing head-on. Here's what I got back:[115]

Dear Ms. Azzoni,

Thank you for contacting me regarding patent reform legislation. I share your concerns and believe that abusive patent litigation is a threat to innovation and costs businesses and consumers billions of dollars each year.

As you know, the TROL Act (H.R. 2045)[116] and the STRONG Patents Act (S. 632)[117] would focus on post-issuance review proceedings at the Patent Trial and Appeal Board in an effort to protect the property rights of the inventors that grow the country's economy. These bills would require the pattern or practice of sending bad faith demand letters to be treated as an unfair or deceptive act or practice in violation of the Federal Trade Commission Act.[118]

---

[115] In her letter, Senator Gillibrand refers to four pieces of legislation that were proposed at the time, and one extant; these will be explained briefly in footnotes within the letter.

[116] "To provide that certain bad faith communications in connection with the assertion of a United States patent are unfair or deceptive acts or practices, and for other purposes." This bill was introduced on April 29, 2015, but it did not receive a vote. It was subsequently reintroduced, most recently on January 5, 2021, in the related bill, H.R. 192. https://www.govtrack.us/congress/bills/114/hr2045

[117] "A bill to strengthen the position of the United States as the world's leading innovator by amending title 35, United States Code, to protect the property rights of the inventors that grow the country's economy." This bill was introduced on March 3, 2015, but it did not receive a vote. As of this writing it has not been reintroduced. https://www.govtrack.us/congress/bills/114/s632

[118] "The Federal Trade Commission Act of 1914 established the Federal Trade Commission. The Act was signed into law by U.S. President Woodrow Wilson in 1914 and outlaws unfair methods of competition and unfair acts or practices that affect commerce." https://en.wikipedia.org/wiki/Federal_Trade_Commission_Act_of_1914

The Innovation Act (H.R. 9)[119] and the PATENT Act (S. 1137)[120] aim to reform the patent system by changing the administrative and judicial processes that protect intellectual property rights. These bills would establish uniform pleading standards, reasonably limit early discovery, and protect end users who are targeted for patent infringement based on their purchase of products from a manufacturer.

Between 2005 and 2014, there have been a total of 2,100 lawsuits against 541 companies filed in New York State alone, with nearly 70% of those lawsuits filed in the past five years. Furthermore, patent litigation rates in the first half of 2015 outpaced the rates in the last six months of 2014 by 32%. The increasing rates of patent litigation show that businesses of all sizes and across all industries are at risk. Please know that I will continue to work with my colleagues to ensure that patent trolls, whose sole purpose is to buy ambiguously worded patents and extract money by suing or demanding licensing fees from companies whose product or technology allegedly infringe upon the claims of the broadly defined patent, are stopped.

Thank you again for writing to express your concerns, and I hope that you keep in touch with my office regarding future legislation and concerns you may have. For more information on this and other important issues, please visit my website at http://gillibrand.senate.gov and sign up for my e-newsletter.

Sincerely,
Kirsten Gillibrand
United States Senator

---

[119] "To amend title 35, United States Code, and the Leahy-Smith America Invents Act to make improvements and technical corrections, and for other purposes." This bill was introduced on June 11, 2015, but did not receive a vote. As of this writing it has not been reintroduced. https://www.govtrack.us/congress/bills/114/hr9

[120] "A bill to amend title 35, United States Code, and the Leahy-Smith America Invents Act to make improvements and technical corrections, and for other purposes." This bill was introduced on June 4, 2015, but it did not receive a vote. As of this writing it has not been reintroduced. https://www.govtrack.us/congress/bills/114/s1137

I read the letter, and then I read it again... and then I passed it along to Paul Morinville. I needed a second opinion, needed someone knowledgeable and intelligent to confirm my suspicion that Senator Gillibrand had the wrong end of the stick, had things totally turned around backwards. Why had she felt compelled to tell me about the plight of businesses receiving "demand letters" and being sued for patent infringement, when I'd just gotten through telling her that that was exactly what I, given the state of things, was forced to (rightfully) do? Was she... in effect... calling me... a *patent troll!?!?* I asked Paul, "Is my read on this correct? That my Senator didn't even get the fact that I'm an NPE only because I can't afford to make my own version of my product, but that I am the *original* patent holder and the only way I can enforce my patents is by sending these demand letters?"

Paul's reply: "You are correct. It's a standard response, and much like the others I've seen. She is regurgitating the language and argument that were given to her."

Well, I'd call that a biiiiiig *swingandamiss*. How could my governmental representative *represent* me if she didn't even understand the situation facing me and hundreds of thousands of indie inventors like me—*even when it was spelled out for her in simple chapter and verse?* When all the straight-from-the-horse's-mouth story got from her was a boilerplate bullshit likely drafted and delivered by some 24-year-old staffer who uptalks and wears flip flops in the Halls of Justice... *ugh.*

Still, hers wasn't the only set of important ears in Washington. A recent piece of legislation—the SUCCESS Act, signed into law by President Trump on October 31, 2018—included an item requiring the USPTO to "study and report to Congress on ways to close the gender, race, and income gap in patenting rates."[121] Accordingly, the USPTO was then conducting their "Study of Under-

---

[121] "H.R. 6758 (115th): SUCCESS Act" https://www.govtrack.us/congress/bills/115/hr6758

represented Classes Chasing Engineering and Science Success,"[122] part of which involved collecting first-hand accounts from inventors in these under-represented classes (defined by Congress as "women, minorities, and veterans") to better understand the particular obstacles and challenges inventors in these underrepresented classes encountered while pursuing patents and navigating the patent system. Husband-and-wife team Jeff Hardin and Patricia Duran—inventor advocates, small business owners, and consultants on IP policy—had just provided their own testimony to the study, and were encouraging others to do the same; the idea was that I, as part of the 8% of indie inventors with two X chromosomes, could provide the USPTO with some valuable insight which they would then provide to Congress.

Thinking it might do some of the good I'd hoped it would do with Senator Gillibrand, I expanded my letter and sent it over.[123] I told how I was the founder and inventor of the CardShark walletskin, a simple solution to a problem we never knew we had. I told how important I felt legislation like the SUCCESS Act to be, not just as an underrepresented inventor but as someone

---

[122] From the USPTO website:
"The Study of Underrepresented Classes Chasing Engineering and Science Success (SUC-CESS) Act of 2018 directed the Under Secretary of Commerce for Intellectual Property and Director of the U.S. Patent and Trademark Office (USPTO), in consultation with the administrator of the Small Business Administration, to prepare a report that:
• Identifies publicly available data on the number of patents annually applied for and obtained by women, minorities, and veterans;
• Identifies publicly available data on the benefits of increasing the number of patents applied for and obtained by women, minorities, and veterans and the small businesses owned by them;
• Provides legislative recommendations for how to promote the participation of women, minorities, and veterans in entrepreneurship activities and increase the number of women, minorities, and veterans who apply for and obtain patents."
https://www.uspto.gov/ip-policy/legislative-resources/successact

[123] The full, 1,900-word letter, along with many other testimonies provided by other inventors who participated in this study, is available to read online at https://usinventor.org/wp-content/uploads/2020/01/SUCCESS-Act-Combined-Written-Testimony-from-24-Witnesses.pdf

who understood the importance of a robust patent system and strong patent protections. I explained that, while I certainly supported any effort to increase the number of women, minority, and veteran patent holders, I personally considered all independent inventors to be a minority in their own class and right: explained that, in my view, the impulse to support "underrepresented classes" within and through the patent system should be broadened to include the entire minority indie inventor demographic.[124] If the USPTO wanted to do this, I said, then it was crucial that they increase patents' enforceability, which had been badly damaged by the advent of the AIA and the PTAB. If the recommendations produced by this study did not include a recommendation for the increased protection of all patents, I said, then any increase in patent holding among these specifically-targeted classes would be meaningless.

My letter, along with the letters of more than 20 other underrepresented inventors and those of Jeff and Patricia themselves, went off to the USPTO on June 30, 2019. On October 31 of that same year, the USPTO's SUCCESS Act report was transmitted to Congress. It's main conclusion? "Overall, there is a need for additional information to determine the participation rates of women, minorities, and veterans in the patent system."

Well, thanks for taking the time. You really know how to make a girl feel special.

Of course, it wasn't just me whose advice they ignored. According to Jeff, over 75% of the comments the USPTO received from all the underrepresented inventors in their study emphasized this same main point: that the inability to enforce a granted patent is a substantial barrier to participation in the patent system in the first place. Jeff and Patricia, in their own speech to the USPTO during the first public hearing of the SUCCESS Act, had already asked point-

---

[124] I wasn't alone with this sentiment; see "Independent Inventors to USPTO: We Are All Underrepresented in This Patent System," Eileen McDermott, IP Watchdog, May 9, 2019. https://www.ipwatchdog.com/2019/05/09/independent-inventors-uspto-underrepresented-patent-system/id=109109/

blank: "What good is a patent if you can't defend it?"[125] Despite all of this, what did the USPTO recommend to Congress as a means of encouraging women, minorities, and veterans to participate in the patent system? A stamp. Yes, you read that right. A stamp. Like, for a letter.[126] The USPTO recommended that Congress "[c]reate a commemorative series of... postage stamps to be placed in circulation... that feature a spectrum of American inventor-patentees from a variety of backgrounds, including those from underrepresented groups."[127]

What's that line? "Isn't it ironic? Like rain on your wedding day? A free ride when you've already paid?" How about: the USPTO thinking that an accessory to *yesterday's* technology is a good way to help those *creating the future*.

Want to hear another verse in this same sad song? Following this report to Congress and its failure, I—along with many other stakeholders who testified to the USPTO—signed a letter informing the House of Representatives Small Business Committee that our comments to the SUCCESS Act report had been all but ignored.[128] I then sent a further expanded version of this same testimony to the Senate's Subcommittee on Intellectual Property of the Committee on the Judiciary for their hearing on "Improving Access and Inclusivity in the Patent System: Unleashing America's Economic Engine." That hearing took place on April 21, 2021, the record came and went, and guess whose statement wasn't read? Not that I was the only one: despite Jeff and Patricia including citations from the testimony provided by me and other underrepresented inventors in their own written statement[129]—a statement in which they also informed

[125] "Duran & Hardin's Oral Testimony - SUCCESS Act." https://www.youtube.com/watch?v=LcofvTs8Beg

[126] You know, those things your grandparents used to send to each other.

[127] "Report to Congress pursuant to P.L. 155-273, the SUCCESS Act," October 2019. https://www.uspto.gov/sites/default/files/documents/USPTOSuccessAct.pdf

[128] Inventor Letter to Congress. https://usinventor.org/wp-content/uploads/2020/01/Inventor-Letter-to-Rep-Velazquez-re-SUCCESS-Act.pdf

[129] https://bit.ly/Hardin-Senate-Improving-Patent-Inclusivity

the Senate what really transpired with the laughable SUCCESS Act study—none of our stories were heard or considered. No wonder talk of politics was taboo at the dinner table when I was growing up: it makes people act out in strange and unpleasant ways. I could hear my mother's voice grown shrill with emotion echoing in my ears, only now it was me who was yelling: "What the hell were you thinking, Kip? What the hell did you think was going to happen?"

Exactly. Why the fuck did I even fucking bother.

What do they say about exercises in futility? They come in threes? Through a mutual acquaintance I managed to get ahold of an email address for Andrei Iancu, the then recently-appointed Director of the USPTO. Will Plut had mentioned that, in a conversation he'd had with Iancu, the new Director expressed a desire to be made aware of any cases of abuse against independent inventors as a result of the PTAB and/or the inter partes review. Thinking of my L.A. experience with the *#metoo* infringer and her husband, I decided to reach out. I drafted an email explaining who I was and what had happened to me in L.A. I explained how the infringer's husband's threat was exactly the sort of procedural abuse that was all but guaranteed by the current legislation. I described the economic realities of my situation: realities that included a financial decision to be made between the expense of manufacturing and the expense of filing and defending the patent itself, a resultant dependence on licensing, and—within the pursuit of this licensing—an inability to secure legal representation on contingency due to the evaporating likelihood of success in the face of an IPR proceeding. I was—as so many indie inventors were—between a rock and a hard place, hung out to dry by the very system that was supposed to protect me. I told him that I would very much like to come to D.C. and share my story in full, to help him better understand the plight of the indie inventor in the post-AIA landscape. I thanked him for his time and I hit send… and I never heard back. Maybe I will one day, though. Maybe he's still getting around to it. I expect he's got a lot more time on his hands, now, since he stepped down from the position of USPTO Director in January of 2021.

Still, maybe Iancu's silence would have been less of a surprise if I'd better understood the recent history of the position he held: if I'd better understood the nature of that position in its present iteration. As I said at the beginning of this section, I'd come to the conclusion that the message had to be delivered to the central players where they lived; I thought that contacting Director Iancu and Senator Gillibrand and the Subcommittees was my best chance to do just that. I thought that if they heard my story—if they understood the real consequences of the recent legislation, no matter how noble its intentions, for the independent inventors, and the threat this implosion represented to America itself—then they would understand the need for immediate and dramatic action. I—naively—believed that the simple and obvious wrongness of the present circumstances, when presented to those with the power to change those circumstances, could and would set the pendulum swinging finally but unimpeachably back. I hadn't fully understood then that it was the state of play of the game itself that now shaped these players' actions: that the game was playing itself, so to speak, running along an architecture built before many of these players even arrived on the scene; that it was the authors of this architecture, now absent from the field, whose influence was still being felt in every PTAB proceeding and IPR threat; that getting to the real heart of the problem would mean trying to understand and disentangle the threads that wove together Big Tech and the USPTO up to and through the transitional period following the passage and implementation of the AIA.

II. THE ANSWER TO 99 OUT OF 100 QUESTIONS... PART 2

Consider the following, taken from the USPTO's "Performance and Accountability Report" from the period including the AIA's passage, Fiscal Year 2011:

The USPTO has evolved into a unique government agency. In 1991— under the Omnibus Budget Reconciliation Act (OBRA) of 1990—the USPTO became fully supported by user fees to fund its operations. In 1999, the American Inventors Protection Act (AIPA) established the USPTO as an agency with performance-based attributes, for example, a clear mission statement, measurable services and a performance measurement system, and known sources of funding. On September 16, 2011, the President signed into law the Leahy-Smith America Invents Act (AIA). The new law will promote innovation and job creation by improving patent quality, clarifying patent rights, reducing the application backlog, and offering effective alternatives to costly patent litigation. It also provides fee-setting authority that will be essential to USPTO's sustainable funding model.[130]

While perhaps unremarkable in itself, this passage takes on a different meaning when considered alongside this passage from AIA-era Director David Kappos's bio page on the USPTO website:

In response to financial pressures resulting from the recession (2007-2009) and a decline in business activity for the agency, the USPTO froze hiring and scaled back programs where possible. The outlook brightened in 2011 when Congress passed the AIA, which Director Kappos helped to shape. He then set to work implementing the Act's provisions, including the establishment of the Patent Trial and Appeal Board, and restarting expansions and initiatives that had been interrupted by the recession.[131]

---

[130] "United States Patent and Trademark Office: Performance and Accountability Report Fiscal Year 2011," USPTO.gov, 2012. https://www.uspto.gov/sites/default/files/about/stratplan/ar/USPTOFY2011PAR.pdf

[131] "David J. Kappos," USPTO.gov. https://www.uspto.gov/about-us/david-j-kappos

...and it takes on a still more interesting aspect when considered alongside the extract below, taken from the USPTO's current (2022) *Fiscal Year Congressional Justification*:

> *The USPTO is a demand-driven, fee-funded, performance-based organization with a commitment to delivering reliable IP protection and information to its various stakeholders... In fiscal year 2022, the USPTO expects to employ 13,723 federal employees, including patent examiners, trademark examining attorneys, computer scientists, attorneys, and administrative staff. Employee engagement is a core component of the Office's business strategy, as it contributes to the recruitment and retention of a diverse, high-performing, nationwide workforce to execute the Office's mission... However, the USPTO routinely assesses its functions and relies heavily on the private sector for those aspects of its operations that are not inherently governmental, such as contracting to third parties the processing of the administrative aspects of the patent and trademark examination processes, as well as certain mission-support activities.[132]*

So for those of you keeping score at home: the Director of a cash-strapped, fee-dependent governmental agency helps to craft a piece of legislation that 1) creates a new, expensive, and sure-to-be-popular means of fee-generation (the PTAB) for his agency, and 2) gives that agency unilateral authority to both set fees and spend fees collected, including to private contractors.

---

[132] "U.S. Patent and Trademark Office: Fiscal Year 2022 Congressional Justification," USPTO.gov, May, 2021. https://www.commerce.gov/sites/default/files/2021-05/fy2022_uspto_congressional_budget_justification.pdf

And also, P.S.: Mission-supported activities like, for example, that oh-so-productive use of my time we're calling the "Study of Underrepresented Classes Chasing Engineering and Science Success"? The one that produced the stunning insight that, "[o]verall, there is a need for additional information to determine the participation rates of women, minorities, and veterans in the patent system"? Makes you wonder who cashed the check for that particular "mission-support activity."

Sound a little murky? Read on.

Getting legislation passed ain't easy. You need a lot of people on your side to make it happen—need to tally a lot of votes in your column. You need a lot of people to see things from your particular point of view, so that when the time comes they vote the right (read: your) way. In such circumstances, it's extraordinarily helpful to have an advocate: someone who's good at getting out there, shaking hands and buying dinners, dispensing "useful" information to those with the power and the authority to make the decisions that affect your legislative future—and your bottom line. If you can understand what I've just described, then you already understand everything you need to know about lobbyists and lobbying. As Investopedia cheerily puts it: "With the number of tasks and matters required of a legislature ever-growing, populaces need lobbying to bring issues front and center, otherwise, the government can fall into an 'out of sight, out of mind' trap."[133]

Of course, as the Director of a government agency, you can't really have lobbyists. You can't use your agency's resources to lobby for legislation on behalf of your agency.[134] What you really need is a "friend on the outside," so to speak... and preferably one with deep pockets. Or, better yet, one with *reeeeeeeally* deep pockets. Someone who can foot the bill for a whole lot of lobbying. And, if not a friend exactly, at least an ally: someone whose interests align with yours. Someone who would like a ride out of the current state of

---

[133] "Why Lobbying Is Legal and Important in the U.S.," Daniel Weiser, Investopedia, July 13, 2021. https://www.investopedia.com/articles/investing/043015/why-lobbying-legal-and-important-us.asp

[134] 18 U.S. Code § 1913 - Lobbying with appropriated moneys - states, in part: "No part of the money appropriated by any enactment of Congress shall, in the absence of express authorization by Congress, be used directly or indirectly to pay for any personal service, advertisement, telegram, telephone, letter, printed or written matter, or other device, intended or designed to influence in any manner a Member of Congress, a jurisdiction, or an official of any government, to favor, adopt, or oppose, by vote or otherwise, any legislation, law, ratification, policy, or appropriation, whether before or after the introduction of any bill, measure, or resolution proposing such legislation, law, ratification, policy, or appropriation."

affairs, off to greener pastures, and who—if you're in a position to help make that happen—wouldn't mind covering the Uber fare if you want to hitch a ride together.[135]

Am I suggesting that David Kappos used Google and Google's lobbying capabilities to pass legislation that opened the cash floodgates for his agency, just as much as Google used him to craft legislation that defanged any would-be litigious patent holder, in an incredibly egregious example of "I'll scratch your back if you scratch mine?" Yeah, I think I am suggesting that. I think I'm suggesting that pretty damn hard.

But it's worse than you think. David Kappos left the position of Director of the USPTO on January 31, 2013. His duties were taken over by Deputy Director Teresa Rea, who herself resigned later that same year, on November 21. The position was then taken up by Commissioner for Patents Margaret A. Focarino, who served in the role until Michelle K. Lee took over the Office on March 12, 2015. And Big Tech? They couldn't have been happier with the appointment.

Don't take my word for it, though. Check out the following, from the USPTO's bio page on our friend Michelle:[136]

> *Michelle K. Lee... was educated at the Massachusetts Institute of Technology, where she earned a Bachelor of Science in electrical engineering and a Master of Science in electrical engineering and computer science, and at Stanford University, where she earned a Doctorate of Jurisprudence.*
>
> *From 1992 to 1994, Lee held successive clerkships in federal courts,*

---

[135] I've already discussed Google's lobbying activities in the period leading up to and including the passage of the AIA in Part II; for the moment, maybe all you need to remember is that Google spent over $11 *million* on lobbyists in 2011, more than twice as much as it spent in any of the previous years (from the Center for Responsive Politics, a 501(c)(3) nonprofit organization that tracks money in American politics through its website, OpenSecrets.org). https://www.opensecrets.org/federal-lobbying/clients/summary?cycle=2013&id=D000022008).

[136] "Michelle K. Lee," USPTO.gov. https://www.uspto.gov/about-us/michelle-k-lee

first under Judge Vaughn R. Walker of the U.S. District Court for the Northern District of California and then under Judge Paul R. Michel of the U.S. Court of Appeals for the Federal Circuit in Washington, D.C. She then went into private practice, specializing in intellectual property (IP).

In 2003, Lee joined Google Inc. as Senior Patent Counsel, eventually attaining the rank of Deputy General Counsel, Head of Patents and Patent Strategy. In recognition of her expertise and experience in the application of IP protections to the technological advances coming out of Silicon Valley, Lee was appointed to the Patent Public Advisory Committee, the body providing guidance to USPTO directors on policy, goals, performance, budget, and user fees.

Lee entered full-time public service in 2012 when she became Director of the USPTO's office in Silicon Valley. Two years later, President Barack Obama appointed Lee to the position of Deputy Under Secretary of Commerce for Intellectual Property and Deputy Director of the USPTO and then nominated her in short order for the agency's top leadership position. The Senate confirmed the nomination, making Lee the first woman to serve as Director of the USPTO when she was sworn in on January 13, 2015.

As Director, Lee prioritized efforts in the areas of information technology (IT), electronic filing (especially with respect to trademarks), patent quality, outreach, and international cooperation.

No need to adjust your screen, you read that right: with Michelle Lee at the head of the USPTO, a former high-ranking Google employee was now running the agency wherein much of Google's legal and financial future would be argued and decided. Sound like a conflict of interest? Sound murkier still? Again, read on:

According to her Huffington Post contributor bio, Lee "joined [Google] when it was relatively young, and was responsible for formulating and implementing its worldwide patent strategy, including building its patent portfolio from a small handful of patents to over 10,500 assets in eight years."[137] She "[l]ed Google's efforts to purchase one of the largest patent portfolios in history (Nortel's for $4.5 billion) via a bankruptcy auction,"[138] "advised the search engine giant on its acquisition of YouTube, [its] participation in the 2009 Nortel patent auction, and on mobile phone patent issues,"[139] and was a "[k]ey part of [mergers and acquisitions'] efforts to value and analyze target company's intellectual property assets and products as well as subsequent integration of acquired technologies and rights."[140] She also "advised executive management on [the] high-stakes, industry-wide smart phone patent wars."[141] She was, to put it mildly, absolutely integral to Google in its rise to industry dominance—and, as we can infer from contemporaneous salary indexes, public records of Google's IPO activities, and Michelle's own published statements, she was well compensated for her efforts.

Contemporaneous estimates place Michelle Lee's likely salary somewhere in the mid six figures—a good gig, certainly, but nothing all that special when we consider what is and was standard pay for someone with Michelle's credentials and experience. Things get a biiiiit more interesting when we consider the other incentives Google is known to have offered to upper-level hires[142] like Michelle: namely stock options in the young and growing company.

---

[137] "Contributor Michelle K. Lee," Huffington Post. https://www.huffpost.com/author/michelle-k-lee

[138] From the description of the job she held at Google, posted on Michelle's LinkedIn profile: https://www.linkedin.com/in/mlee95070/

[139] "Director of the U.S. Patent and Trademark Office: Who Was Michelle Lee?" Steve Straehley, AllGov.com, June 12, 2017. http://www.allgov.com/news/top-stories/director-of-the-us-patent-and-trademark-office-who-was-michelle-lee-170612?news=860210

[140] From the description of the job she held at Google, posted on Michelle's LinkedIn profile: https://www.linkedin.com/in/mlee95070/

[141] *Ibid.*

[142] I use the term "upper-level" with some confidence: Google's masseuse was known to have received stock options as part of her compensation package, enough to make her a millionaire after Google's IPO, so it's no small leap to assume a desirable arrangement for someone in Michelle's highly-valued position within the company.

For those of you who've never been hired as an executive at a major corporation: a "stock option is the right to buy a specific number of shares of company stock at a pre-set price, known as the 'exercise' or 'strike price,' for a fixed period of time, usually following a predetermined waiting period, called the 'vesting period.'"[143] As a new hire in 2003, Michelle would have received an estimated 10,000[144] options. When Google filed its IPO on August 19, 2004 it did so with an offered share price of $85. Though we don't know the strike price Michelle was offered, we can reasonably assume it was in the ballpark of the IPO price, if not better: we know that, as PersonalCapital.com states, "[w]hen employees are given stock options at an early-stage startup, they usually have the right to buy shares at a very low valuation." Go with me on the $85 per share price, then, and I think you'll see why that number less important than the next one: at the close of the stock market on January 13, 2014, the day Michelle Lee assumed the role of USPTO Director, Google's stock closed at $559.30 a share.

OK, so she did well. Good for her, right? If she exercised her options and then sold that stock and got out of the company before taking on her role as USPTO Director she would have walked away a very wealthy lady, but that was just her good luck. Right place at the right time, right? Which she must have done. Because she couldn't very well hold a stake in a company whose fortunes she would be in a position to affect as the head of an (allegedly) impartial governmental agency, right? She got rid of the stock, right? Guys? Right?

Certainly, at the very least, that's the sort of thing that somebody would check. And, as it turns out, they do: according to the United States Office of Government

---

[143] "Stock Options: What Are They and How Do They Work?" Personal Capital, May 17, 2021. https://www.personalcapital.com/blog/investing-markets/how-stock-options-work/

[144] There is debate about this number, though it is important to note that this is a conservative estimate. See: "Google Faces Brain Drain As Anniversaries Hit," Sam Savage, RedOrbit.com, April 11, 2007. https://www.redorbit.com/news/technology/898498/google_faces_brain_drain_as_anniversaries_hit/index.html/

Ethics (OGE), "[t]ransparency is a critical part of government ethics, and Congress has determined that the citizens should know their leaders' financial interests. To facilitate such transparency, Congress enacted the financial disclosure provisions of the Ethics in Government Act. The Act imposes detailed requirements for public financial disclosure by senior United States Government officials. The OGE Form 278e and the OGE Form 278-T are financial disclosure reports that request only as much information as the Act requires a filer to disclose."[145] And, naturally, during her confirmation process, our friend Michelle dutifully filled out this form. And what, pray tell, did she disclose?

The form, submitted to the Department of Commerce on September 9, 2014, reported personal assets with a "minimum value" of $8,956,041.00 and a maximum value of $26,659,000.00. Much of the value of these assets exists within the "Michelle K. Lee Living Trust" based in Saratoga, California, and established in September, 2007. California rules governing a living trust[146] state that "[a] trust is not a public record.... So, the general public or anyone who is not a beneficiary does not have a right to know about the assets in your trust." Meaning that, as the sole trustee on the account, Michelle is not required to publicly state the contents of the trust.

So... to quote Brad Pitt's character in *Se7en*... "*WHAT'S IN THE BOX?!?!*"

When asked, on the Senate Committee on the Judiciary's "Questionnaire for Non-Judicial Nominees," to "[i]dentify... financial arrangements that are likely to present potential conflicts-of-interest when you first assume the position to which you have been nominated," Lee responded simply, "I am not currently aware of any potential conflicts of interest."[147] Reporter Klint Finley, comment-

---

[145] "Public Financial Disclosure Guide," United States Office of Government Ethics. https://www.oge.gov/Web/278eGuide.nsf

[146] "Living Trusts," The Superior Court of California: County of Santa Clara. http://www.scscourt.org/self_help/probate/medical/living_trust.shtml#what

[147] "United States Senate Committee on the Judiciary: Questionnaire for Non-Judicial Nominees for Michelle Kwok Lee," Judiciary.Senate.gov, October 28, 2014. https://www.judiciary.senate.gov/imo/media/doc/Lee%20Questionnaire%20Final.pdf

ing on Lee's appointment for Wired.com, wrote, "While it's good news that someone who understands the problems with the modern patent system will likely be head of the USPTO, it's hard not to worry that Lee is too close to the technology industry, and Google in particular."[148]

Yeah, no shit. Understatement of the year, Klint. Understatement of the millennium.

Still, perhaps we're being unfair—judging the book by its cover, so to speak. After all: just because she worked for Google doesn't mean she'd thumb the scales for their benefit. And we can't assume, just because her assets are arranged in a particularly and conveniently obfuscatory way, that this means some sort of intentional deception or collusion. Plenty of people have living trusts, and we certainly can't assume that hers is or ever was full of Google stock, giving her a substantial stake in the company's fortunes. We should hesitate, too, to leap to any conclusions about the fact that, throughout the entire vetting process, Michelle's soon-to-be boss—Secretary of Commerce Penny Pritzker—was just a phone call away: Penny Pritzker, whose own financial gymnastics were later revealed in the infamous Paradise Papers; Penny Pritzker who, in advance of her own appointment, "divested" herself of conflict-of-interest assets… by stashing them in offshore tax havens.[149] No, let's give our friend Michelle the benefit of the doubt. Let's judge her as we'd like to be

---

[148] "Silicon Valley Wins! Obama Picks Ex-Google Lawyer to Head Patent Office," Klint Finley, Wired.com, October 17, 2014. https://www.wired.com/2014/10/michelle-lee-uspto/

[149] See the article "Former Secretary of Commerce And Hyatt Hotels Heir Penny Pritzker Identified In Paradise Papers" Angel Au-Yeung, Forbes.com, November 5, 2017, for more of the story. https://www.forbes.com/sites/angelauyeung/2017/11/05/former-secretary-of-commerce-and-hyatt-hotels-heir-penny-pritzker-identified-in-paradise-papers/?sh=a2f31242618e. Also, to be clear: Michelle Lee is not named in the Paradise Papers, and there is no direct documentation showing that she engaged in any illegal—or even unethical—financial activity. Still: it kind of makes you wonder about the OGE's request for "only as much information" as the Act requires a filer to disclose, doesn't it? It makes you wonder about those requirements, and about who came up with them, and about what, exactly, they leave out.

judged: on the merits of her actions. I'm sure that Lee's USPTO treated Google as it would any other entity involved in an IP dispute. I'm sure there's no evidence of bias at all.

...except you already know that isn't how it went down. You know because you've already met my good friend, David Hoyle. Remember David and his company B.E. Technologies, from back in Parts II and III?[150] It turns out that David's story is much more than just a story about the PTAB being wielded as a weapon by Big Tech infringers against little guy infringees; it turns out it's a story about how the USPTO had become, in effect, a weapon that was now wielding itself.

Before we proceed, I need to tell you that what follows is deeply, deeply indebted to the work of both David and his attorney, Agatha M. Cole, upon whose research and reconstruction of events the following heavily relies.[151]

Back in the mid-'90s, if you used a personal computer at work and another one at home, you ran very quickly up against the limitations of the data-sharing technology of the time. Specifically, there was no good way to access information stored on one machine from another—no good way to work on a project on your computer at work, continue working on that same project file at home, and then complete your work on that file back at the office in the morning. There was a lag between the technology that existed and the technology that was needed, and David Hoyle—then working as a computer programmer—had an idea for how to fill the gap. Better than that, he had an idea for how to provide his products free-of-charge to end users, subsidizing the costs with advertising revenues in a new and powerful way. Realizing that the nature of the internet—particularly the in-

---

[150] If you need a refresher: David invented and held patents on software the Google wanted: they infringed, he sued, they turned around and IPR'd him, and the PTAB invalidated David's patents; this despite the fact that David's patents were filed and validated, and his software was operating in the space, years before Google even arrived on the scene.

[151] https://patentlyo.com/media/2021/08/Hoyle-v.-Lee-Complaint.pdf

formation gleaned by the user's connection to it and activity on it—offered a unique opportunity to both discover a user's needs and interests and provide personalized advertising targeting those needs and interests by means of web-based applications, David set about to design technologies to do just that, founding B.E. Technologies in August 1997 to pursue the idea.

In the years that followed David applied for and obtained a total of 10 patents pertaining to these web-based technologies, most notable among these those we mentioned back in Part II: U.S. Patent No. 6,628,314 (what we'll call the "314" patent for short), filed on July 17, 1998 and issued on September 30, 2003, and U.S. Patent No. 6,771,290 (the "290" patent), also filed on July 17, 1998 and issued on August 3, 2004.

The "314" patent concerns the technical aspects of technologies to gather user-specific data for the purpose of delivering user-relevant advertising via various online applications; the "290" patent concerns the technical aspects of personalization features for internet-based web browsers, including a toolbar that provides the user with icons to access programs, files, and an internet search; this toolbar, with the user's consent, gathered data that was used to deliver targeted advertising products via various online applications.

In the years following the issuance of these patents, between late 2006 and 2007, the USPTO rejected several patent applications filed by Google on the grounds that the targeted advertising technologies described in their applications had already been patented in David's "314" and "290" patents. Tellingly, rather than amend its applications in hopes of gaining protection in a narrower or more distinct invention—the standard play for an applicant whose initial application is rejected—Google instead abandoned all of the patent applications relating to the targeted advertising claims covered by David's patents. They didn't even try to skirt the hurdle: they just straight-up ignored it.

And who was at the Head of Patents and Patent Strategy at Google at this time? Of course it was our good friend, Michelle Lee.

In early 2007, David became aware of the fact that Google—along with several other industry-leading technology companies—had implemented targeted advertising technologies that directly infringed on his IP. Accordingly, on September 7, 2012, B.E. Technologies filed patent infringement legal action against several of these infringers, Google, Facebook, Microsoft, and Samsung among them.

Of course we all know what else happened in 2012: while Google was busy spending a record $18 million on federal government lobbying—much of it directed towards intellectual property reform—the USPTO itself began to implement the changes dictated by the newly-passed America Invents Act, most notable among these the creation of the IPR and the PTAB whose Administrative Patent Judges (APJs) oversee those reviews. These APJs, remember, are appointed by the Secretary of Commerce in consultation with the Director of the USPTO; remember also, as we move into the next part of our story, that a panel of three APJs presides over each IPR proceeding.

As we've discussed, under the new, post-AIA system, any party can file a petition to institute an IPR against any other party, challenging the validity of that party's existing patent; practically, however, as we've also discussed, the high costs associated with filing an IPR generally make this course of action feasible only for those with sufficient capital reserves to support it. As Josh Malone, in his Law360 article "PTAB Trials Disproportionately Harm Small Businesses," says: "It is rare that a small business will avail itself of the PTAB to fend off a meritless patent assertion... Conversely, it is common for a large corporation to weaponize the PTAB to crush a smaller competitor with superior technology... Large corporations and patent assertion entities can absorb the expense and risk of PTAB trials as a cost of doing business. They have access to billions of dollars in capital and portfolios of hundreds or thousands of patents. They need only factor in the cost and risk of PTAB invalidation in their budget and business plans."[152] It's no surprise, then, that the majority of IPR

---

[152] "PTAB Trials Disproportionately Harm Small Businesses," Josh Malone, Law360, January 29, 2021. https://www.law360.com/articles/1348182/ptab-trials-disproportionately-harm-small- businesses?copied=1

petitions are filed by large, deep-pocketed parties—such as tech Goliaths like Google—against small indie inventors like David... and no surprise that, with access to the remedies now available under the new system, Google and the other defendants in David's suit filed petitions to challenge the validity of his "314" and "290" patents. Their motion to stay the proceedings of David's suit against them until the PTAB had rendered a decision—a motion which was granted—was just the icing on the cake.

Over the coming months, the PTAB conducted seven separate IPR proceedings concerning the validity of David's patents. All seven were assigned to the same three-judge panel: Sally C. Medley, Kalyan K. Deshpande, and Lynne E. Pettigrew. In the end, the panel ruled against David and B.E. Technologies in each of these seven cases, invalidating both patents as either "anticipated by" and/or "obvious," in light of prior art. David appealed these decisions to the U.S. Court of Appeals for the Federal Circuit; the Court combined the cases into two consolidated appeals, one for all of the decisions concerning each of the patents, and then affirmed the PTAB's ruling.

David's patents—and his lawsuit against the Big Tech infringers—were dead.[153]

But the story isn't over just yet. In mid-2017 a variety of news outlets, blogs, and websites began publishing new revelations about the USPTO's inner workings which, to David, raised serious questions about the constitutionality of his PTAB experience—and of the IPR proceeding itself. In August of that year a government attorney for the USPTO admitted that the agency's leadership routinely engaged in the practice of "stacking" PTAB panels with specific APJs to ensure that PTAB decisions would align with various unwritten policy objectives. Gene Quinn, in an article titled "USPTO Admits to Stacking PTAB Panels

---

[153] Not mad yet? Let me hit you with an interesting tidbit of history, just for context. David filed his patents on July 17, 1998. Do you happen to know when Google was founded? You don't have to Google it, I'll tell you. The company was founded in Menlo Park, CA, on September 4, 1998. You read that right. When David filed his patents— when he staked his claim to the IP Google was now using as though it was their own right and due—*that company didn't even exist.*

to Achieve Desired Outcomes" and published on IPWatchdog.com,[154] quotes the following exchange between the USPTO's attorney and Judge Taranto during the case Yissum Research Development Co. v. Sony Corp.[155]:

Judge Taranto: *And any time there has been a seeming other-outlier, you've engaged the power to reconfigure the panel so as to get the result you want?*

USPTO: *Yes, your Honor.*

Judge Taranto: *And, you don't see a problem with that?*

USPTO: *Your Honor, the Director is trying to ensure that her policy position is being enforced by the panels.*

"In other words," says Quinn, "the Director stacks PTAB panels with Judges that are known to hold views on issues in alignment with the Director."

Certainly this revelation would be cause for concern on its own, but it turns out that it gets much worse. In mid-2020 reports emerged about a previously unpublicized and highly problematic system of incentives operating within the USPTO for APJs appointed to the PTAB. While an APJ's base salary itself is subject to a statutory cap,[156] the USPTO had created a workaround, implementing a system for allocating bonus payments based on an internal points system awarding "decisional units" according to the type and quality of work that an

---

[154] "USPTO Admits to Stacking PTAB Panels to Achieve Desired Outcomes," Gene Quinn, IPWatchdog.com, August 23, 2017. https://www.ipwatchdog.com/2017/08/23/uspto-admits-stacking-ptab-panels-achieve-desired-outcomes/id=87206/

[155] YISSUM RESEARCH DEVELOPMENT CO v. SONY CORPORATION , No. 15-1342 (Fed. Cir. 2015). https://law.justia.com/cases/federal/appellate-courts/cafc/15-1342/15-1342-2015-12-11.html

[156] See 35 U.S.C. § 3(b)(6): "The Director may fix the rate of basic pay for [APJs] ... at [a rate] not greater than the rate of basic pay payable for level III of the Executive Schedule."

APJ undertook and completed within a calendar year. This points system included assessments of such intangibles as the "quality" of an APJ's written opinions, their interactions with USPTO "stakeholders," and their level of support for the PTAB's mission and leadership.[157] According to Steve Brachmann and his article "Financial Incentive Structure for AIA Trials Destroys Due Process at PTAB, New Vision Gaming Argues": "each APJ rated as 'Fully Successful' is eligible for financial bonuses ranging from $4,000 to $10,000, as well as a potential 5% salary bonus."

Remember: following the enactment of the AIA, the USPTO's operating budget—and the availability and amount of any such bonuses—is wholly dependent on the fees that the USPTO both sets and collects; it should go without saying, then, that this bonus system created a particularly problematic set of incentives in the context of the IPR, which had come to generate a substantial amount of revenue—approximately $42,000 in filing fees alone per filing—for the Office. Consider especially the fact that each IPR begins with a petition from the filer which the APJs must assess before deciding whether to institute the proceeding: this assessment, intended to rely solely on the merits of the petition, instead takes place within the context of the strong financial incentive for the Office to grant institution *regardless of the underlying merits of the case.*

But it's worse than that. Subsequent revelations[158] showed that USPTO leadership had been effectively manipulating the "performance-based" components of this compensation structure to control and coerce specific outcomes by implementing policies that expressly discouraged and penalized the issuance of dissenting opinions. In an internal email communication made pub-

---

[157] These revelations originated with documents produced by the USPTO in response to a Freedom of Information Act request filed by U.S. Inventor; a compilation of these documents is available online at https://usinventor.org/wp-content/uploads/2020/05/FOIA-F-19-00277-2019-11-04-APJ-PAPS.pdf.

[158] See "Structural Bias at the PTAB: No Dissent Desired," Gene Quinn, IPWatchdog.com, June 6, 2018. https://www.ipwatchdog.com/2018/06/06/structural-bias-ptab-no-dissent-desired/id=94507/

lic through a FOIA request and addressed to all APJs, Vice-Chief Administrative Patent Judge of the PTAB James T. Moore wrote that dissenting and concurring opinions "are not normally efficient mechanisms for securing the 'just, speedy, and inexpensive' resolution" of PTAB cases, and would therefore no longer be counted towards APJs' "productivity goals."[159] We have no need to wonder then, why approximately 98% of the decisions issued by the PTAB as of 2017 were unanimous;[160] for APJs, the money was in *proactive* affirmation. Note also that by tweaking its incentive system in this specific way, the USPTO effectively turned the three-APJ IPR process into a one-man or -woman show: a single APJ, often hand-selected by the Director of the USPTO in order to achieve a specific pre-determined outcome, would render his or her opinion, and the other two APJs, with their own "decisional units" in mind, would go along.

And once more, for the cheap seats: who was the Director of the USPTO while all of this was going on? Of course it was our good friend Michelle Lee, former Head of Patents and Patent Strategy at Google.

In light of these new revelations, David started doing some digging of his own. He learned that the three APJs who had presided over his own IPR—Deshpande, Medley, and Pettigrew—were known to have a particularly high tendency to rule against patent owners in IPR proceedings as compared to other PTAB judges. In fact, as of 2015, APJ Medley had ruled in favor of the petitioners by cancelling the challenged patents in 100% of the 64 IPR proceedings over which she had presided. APJ Deshpande had likewise cancelled the challenged patents in 100% of the IPR proceedings over which he had presided as of 2015. Luckily, the tribunal was balanced by APJ Pettigrew... who, as of 2015, had cancelled only 97% of the patents that were challenged before her.

At the time of David's IPR there were more that 200 APJs employed by the USPTO, and one has to wonder whether any other collection of three APJs would have had a more predictable response to the matter at hand.

---

[159] *Ibid.*

[160] "Judicial Independence & The PTAB: The Tension Between Judicial Independence & Agency Consistency," Scott McKeown, Ropes & Gray, December 12, 2017. https://www.patentspostgrant.com/judicial-independence-ptab/

But there's still more. David learned that all three of the APJs assigned to his IPR proceeding received substantial bonus payments the year of his IPR. Specifically APJ Deshpande, who wrote the final decisions cancelling the "314" patent, received a bonus payment of $25,220 that year. APJ Pettigrew, who wrote the final decisions invalidating the "290" patent, took home a bonus in the amount of $18,520. APJ Medley, who joined in each of those decisions, received $25,976.

But it gets worse still. David soon uncovered additional facts suggesting that USPTO leadership had *previously and specifically* targeted B.E. Technologies for increased scrutiny and (potentially) adverse actions. He learned that some of his other patent applications had been flagged for enhanced review and scrutiny under the USPTO's Sensitive Application Warning System (SAWS) program, which we discussed briefly in Part II. You'll recall that, during the tenure of this program, the USPTO flagged applications filed by independent inventors that were perceived to potentially threaten existing powerful firms and industries, often rejecting those applications after substantial delay. The assistant patent examiner who had been assigned to David's patent applications actually admitted, during a conversation with a private investigator in July of 2020, that she was instructed to flag all patent applications relating to targeted advertising technologies—including several filed by David—for increased scrutiny under the SAWS program during her employment at the USPTO. Shocking, yes, but—given what we know now—hardly surprising: at the time, targeted advertising comprised almost all of Google's gross revenues. These patents, if issued, would have stood between Google and the profits it wanted and, in reality, depended on to support its rapid expansion; pushing the USPTO to slow these patents' processing and, in some cases, outright prevent their approval meant untold sums saved in licensing and litigation.

To put it simply, it was clear that Google had its eye on David long before the IPR that stripped his patents; it is little wonder, then, that in a system designed to take three years on the outside, David's "314" and "290" applications spent five and six years, respectively, in processing. One wonders if, somewhere in the bowels of Google's reams of documents, is a scrap of paper calculating the margin on that particular delay.

I'd like to jump in here for a second and reiterate an important point about the hub of this whole wheel, namely *money*. The AIA gave the USPTO the unilateral authority to set its own fees and dispense the monies collected as it sees fit— and this includes issuing them as "performance-based" bonuses to APJs. It is worth considering, then, Google's value to the USPTO as a deep-pocketed *customer* that didn't mind spending its cash: a deep-pocketed customer that doesn't quibble over a few thousand—or hundred thousand—here and there. Publicly-available documents show that, as of this writing, Google has filed nearly 400 (396) IPR petitions—placing it in the number three spot for most petitions filed to date, after only Samsung and Apple (who, we should note, were also petitioners in David's IPR).[161] We have to consider, then, the strong incentive the USPTO has to provide satisfactory results for those "stakeholders"[162] that generate the lion's

---

[161] Top 20 Petitioners (PTAB Search: Analytics), UNIFIED PATENTS, https://portal.unified-patents.com/ptab/analytics/case-level/top-parties?sort=-filing_date (aggregating publicly-available data on PTAB cases from the USPTO).

[162] In statements and testimony given during her term as Director, Michelle Lee refers often to the USPTO's "stakeholders." While the term itself is fairly innocuous, and is used throughout the Office to refer generally to—as is stated on the USPTO's "Office of Stakeholder Outreach and Patents Ombudsman (OSOPO)" page—"inventors, patent applicants, and attorneys," it somehow feels different coming from Michelle. Take the comments she made during a Patent Public Advisory Committee Meeting in Alexandria, Virginia on August 14, 2014:

"...From the efforts that led to the America Invents Act, to the process by which the agency exercised its fee setting authority for the first time, we've stood together to advance the interests of this agency and our stakeholders."

"...I'll just conclude by saying how pleased I am that you're all here and that we are all working together under the many important initiatives before the PTO and so important to our stakeholder community."

"...Let me just say that all of my actions here at the PTO are guided by the number one leadership of the agency providing strategic direction to the agency and to the 12,000 hardworking men and women who work here every day. And also to provide transparency and engagement with our stakeholders, whether that be in our operations, whether that be in our performance, our backlog, and our pendency, whether it be in our rule-making or the issuance of our new guidelines."

Maybe it's just me, but given the context of Michelle's deep connection to Google it's hard not to hear that connection in each use of the word: hard not to imagine that, coming from her, the word "stakeholder" refers to the Office's best customers: the handful of deep-pocketed Tech Giants, of which Google is a central member.

You can read the full transcript of the Patent Public Advisory Committee Meeting at https://www.uspto.gov/sites/default/files /documents/ppac_transcript_20140814.pdf

share of its revenues—and its corresponding antipathy towards indie upstarts like David who throw a wrench in those customers' plans.

But there's still more. Only recently, in July of 2021, David became aware of a paper that was published in draft form on the Social Science Research Network (SSRN) which contained several additional troubling revelations about the inner workings of the PTAB at the time of his IPR. The paper, written by Ron D. Katznelson, PhD, and titled "The Pecuniary Interests of PTAB Judges: Empirical Analysis Relating Bonus Awards to Decisions in AIA Trials,"[163] "found that in fiscal year 2016, PTAB judges involved in AIA trials earned a median of more than 14% of their base pay in bonus awards tied to their role as adjudicators... In addition... APJs appeared to have earned an average bonus of $255 per decision when granting [IPR] institution... They also appeared to have earned an average bonus award of $314 per Final Written Decision when canceling patent claims, but only an average of $2 per Final Written Decision when upholding all patent claims." The paper also discusses an APJ recruitment brochure, published by the USPTO, advertising the availability of "gain-sharing bonuses" as a benefit available to PTAB judges—suggesting that APJs would receive a share in the revenues they generated for the agency *by instituting IPR proceedings.*[164]

Taken together, these facts—the USPTO's practice of stacking panels to achieve specific outcomes, Google's inside connection to USPTO leadership during David's IPR proceedings, the system in place to incentivize APJs to institute IPR proceedings, the documented correlation between APJ bonus payments and adverse outcomes for patent holders, the ways in which this compensation structure has enabled USPTO leadership to exert undue influence over the IPR process behind closed doors, and the agency's demonstrated bias

---

[163] "The Pecuniary Interests of PTAB Judges: Empirical Analysis Relating Bonus Awards to Decisions in AIA Trials," Ron D. Katznelson, PhD, Social Science Research Network, July 5, 2021. https://ssrn.com/abstract=3871108.

[164] *Ibid.*

against independent inventors, particularly in the context of targeted advertising, as demonstrated by the existence and implementation of the SAWS program—painted a clear picture: Michelle Lee had employed her case-assignment authority to staff David's IPR proceeding with APJ panelists sure to invalidate his "314" and "290" patents, irrespective of the facts presented. Within Michelle Lee's USPTO, David never stood a chance.

As of this writing—August, 2021—David is pursuing litigation against Michelle Lee and the current and former APJs named in the preceding. The lawsuit, filed in the United States District Court for the Western District of Tennessee, the District covering B.E. Technologies' principal place of business in Memphis, Tennessee, alleges that—according to the preceding—the patent Office's handling of Google's challenges to his advertising technology patents violated David's constitutional due process rights. Time will tell whether justice will be served, or whether the heavy push of Big Tech's influence will once again carry the day. I'm crossing my fingers... but I'd be lying if I told you I was holding my breath.

III. ALL TOMORROW'S PARTIES

It was generally assumed that President Trump would retain Michelle Lee Director of the USPTO; however, her future with the Office became increasingly unclear in the weeks and months that followed his inauguration, as no official statement was issued on the topic from either the White House or the USPTO. Several tech giants—Amazon, Facebook, Samsung, and (naturally) Google among them—sensing that Lee's position was, perhaps, a tenuous one, voiced defenses of her tenure as Director, and publicly pressured President Trump to renominate her. However, on June 6, 2017, Lee resigned. The resignation was sudden and unexpected, and was done without much explanation—though one wonders if lucrative opportunities in the private sector were a motivating factor. As of this writing Michelle is serving as Vice President of the Machine Learning Solutions Lab at another tech industry leader—and prominent defendant in David Hoyle's infringement lawsuit—the tech Goliath Amazon.

Following Michelle's departure, USPTO Associate Solicitor Joseph Matal took over the role and performed the functions and duties of Director until Andrei Iancu was sworn into the position on February 8, 2018. Iancu's tenure as Director was, by many accounts, a step in the right direction for the Office: according to a Law360 article,[165] he "made numerous changes at the Patent Trial and Appeal Board that... increase the chances that patents will survive challenges." The notion that a PTAB proceeding meant certain death for a patent was "definitely on Director Iancu's mind," said James Cleland, an associate at the law firm of Dickinson Wright PLLC, who is quoted in the Law360 article, and it "was pretty clearly a trend he wanted to reverse."[166] As part of this effort, the USPTO under Iancu's leadership "implemented rules changing the claim construction standard for the reviews so that it aligned with the district court's, making it easier to amend claims and harder to mount multiple challenges on the same patent, and letting the board exercise its discretion not to review patents when a trial is looming in district court."[167] That has "led to the board instituting review of fewer patents and upholding more of the patents it does review, a development that has been welcomed by patent owners but"—notably—"knocked by tech companies."[168]

---

[165] "Iancu Leaves Pro-Patentee Legacy As USPTO Director," Ryan Davis, Law360, January 21, 2021. https://www.law360.com/articles/1347266/iancu-leaves-pro-patentee-legacy-as-uspto-director

[166] *Ibid.*

[167] *Ibid.*

[168] *Ibid.*

As I mentioned previously, Iancu left the position in January of 2021, at the end of Donald Trump's presidency. In an address given shortly before his departure, he called on members of Congress to pass laws strengthening the patent system by, among other things, clarifying eligibility standards: if they didn't, he said, we "risk[ed] our nation being left behind as [other nations] fortify their IP laws."[169] As of this writing President Joe Biden has yet to nominate a replacement but, "[w]hen that happens, attorneys will be watching with interest to see if Iancu's policies remain in place or if the next director arrives with a different philosophy."[170] As Gene Quinn says, "[g]iven the support the Biden-Harris ticket received from Silicon Valley[171]... legitimate fears are percolating about whether a first Biden term might look an awful lot like a third Obama term insofar as patents and innovation policy is concerned."[172] With so many of Iancu's policy changes favoring patent owners, James Cleland said, "I would not be surprised to see that... the pendulum swings a little bit back... on those issues."[173]

---

[169] *Ibid.*

[170] *Ibid.*

[171] According to an article published on Observer.com, donations from Big Tech companies, "their employees and their CEOs... soared past $50 million this election cycle." The biggest recipients? Joe Biden and Democratic super PACs. See "Big Tech and CEOs Poured Millions Into The Election. Here's Who They Supported" by Sissi Cao and Jordan Zakarin, published November 2, 2020 on Observer.com, for details. https://observer.com/2020/11/big-tech-2020-presidential-election-donation-breakdown-ranking/

[172] "Facing the Consequences: Biden's Transition Team Should Concern the IP Community," Gene Quinn, IPWatchdog.com, November 12, 2020. https://www.ipwatchdog.com/2020/11/12/facing-consequences-bidens-transition-team-concern-ip-community/id=127279/

[173] "Iancu Leaves Pro-Patentee Legacy As USPTO Director," Ryan Davis, Law360, January 21, 2021. https://www.law360.com/articles/1347266/iancu-leaves-pro-patentee-legacy-as-uspto-director

PART V

LET ME SEE YOUR WAR FACE

"Without work all life goes rotten."
—Albert Camus

"AYAYAYAYAYAYAYAY WHERE DO WE GO NOW?"
—Guns N' Roses, "Sweet Child O' Mine"

On August 11, 2017, a group of indie inventors assembled on the steps of the United States Patent and Trademark Office in Washington, D.C. The gathering, organized by U.S. Inventor, was to protest the high rate of patent invalidations coming out of IPR proceedings before the PTAB. Standing there in the mid-August heat, in full view of the USPTO, these indie inventors took out their patents, dumped them in a pile, and set them on fire. To those watching,[174] the message was clear: in today's patent system, the government's signed and

---

[174] And I do mean to *those* watching. Multiple major news outlets—BBC, FOX, and C-SPAN—covered the event, and Acting Director Joseph Matal put in a brief appearance when the group first assembled, but no one from the USPTO itself was actually there to watch when those patents went up in flames.

sealed assurance of IP protection wasn't worth the parchment it was written on. In today's upside-down IP world, the government's assurance wasn't worth a thing.

Of course, at this point, I'm telling you something you already know. You already know that the patent system is in a sorry state in this country: already understand how Big Tech threw their weight and their influence around to fundamentally upend the very nature of this Constitutionally-enshrined right. You already get how, with each passing year, the corrosive effect of the legislation passed in 2011 grows more pronounced, dropping us further and further down the international IP systems totem pole. The question, as my man Axl so eloquently put it, is: "Where do we go *now?*"

I. BUNGLE IN THE JUNGLE

Before we get into that, though, there's another—and potentially more vital—question we have to ask, and one I'm sure all you indies and aspiring indies reading this are already asking. After all, in the IP jungle as in the real jungle: it's great to know where you're going, but that information won't do you any good if *you don't know how to survive long enough to actually get there.* We've covered a lot of the whats, whys, hows, and whos of the recent changes to the patent system; now, as we enter into the last phase of our journey together, I want to leave you with as much good ol' practical advice as I can about how to cover your inventions, your patents, and 'yer ass. There are a lot of ways to die out here in the IP jungle—a lot of ways to fall prey to the predators, to go broke and lose hope, to "tune in, turn on, and drop out" back to the rat race and the 9-to-5—and, if this book does nothing else, my one hope is that it helps you see, understand, and avoid some of these traps: that it gives you the knowledge and skills you need to steer clear of the potholes and pitfalls that, more than once, almost sent me over the handlebars and skidding across the highway.

Considering how best to tackle this subject got me thinking about my buddy Tim Larkin. Tim is a former military contractor and is the author of the books *When Violence is the Answer: Learning How to Do What It Takes When Your Life Is at Stake, How to Survive the Most Critical 5 Seconds of Your Life,* and *Survive the Unthinkable: A Total Guide to Women's Self-Protection*; he's also a highly sought-out speaker and self-defense instructor.[175] In his classes and seminars, Tim teaches students how to protect themselves from the truly violent predators in society: teaches them how to think strategically, how to quantify and assess their strengths and vulnerabilities, and how to anticipate and avoid the situations that might result in them coming to harm at the hands of an assailant. Tim's point-by-point instruction and distillation of the nature of risk, violence, and predation—honed by decades of training, analysis, and hands-on experience—cuts through the noise and the nonsense and delivers plain, simple, bedrock truths and real, practical tools that attendees can internalize and walk away with. I figure that, if I'm going to try to teach you how to protect yourself in a jungle full of patent predators, I can do no better than to follow Tim's example. Accordingly, here's a list of terms, concepts, outright warnings, and concrete tools that—if I do my job right—will help you navigate the crazy world you're stepping into, and not lose your ass in the process.

Situational awareness. When I met Tim I'd just returned stateside after a several-month stint deep in the Amazon jungle. I'd been living with an indigenous tribe, the Yanomami, while researching the story of illegal gold-mining in the Roraima region nearby. I was still buzzing from the experience, and I quickly rattled off a few of the more colorful moments: the time my bodyguard, leaning over the boat engine's open gas tank with a lit cigarette dangling from his

---

[175] You can learn more about Tim and what he teaches on his website, www.TimLarkin.com.

lip, worked some sort of swap with the locals, gunpowder for bullets; the night the chieftain's wife slipped off with the chief of the next tribe over, sending the village into a violent uproar.[176] Tim seemed to be listening, but I could see his wheels turning: later I understood that, as I spoke, he was quickly and quietly assessing all of the ways that things could have played out differently and turned out very, very badly for me. At this point it was just second nature: his Spec Ops training had taught him that, in extreme situations, things tend to go sideways, that for most people "it's the scenario [they] didn't imagine that unravels everything,"[177] and that good self-defense means, more than anything, good anticipation. There was plenty I hadn't imagined or anticipated when I set off into the jungle, and my sitting back in the States was not proof that I'd made all the right moves: sitting there with Tim and listening to him explain his background and a few of his thoughts on what I'd experienced, it struck me that I'd been very, very lucky.

All of which brings us neatly to point number one: situational awareness. In the jungle or in the IP jungle, the level to which you observe and assess (and reassess, and reassess…) your situation and surroundings and the potential risks and opportunities they present will determine everything about the quality of the decisions you make going forward. Out in the jungle this may mean everything from knowing what the weather is doing to knowing why it's better to sleep off the ground; in the IP jungle this can mean everything from knowing your market segment to knowing your infringer and his or her litigation

---

[176] Middle of the jungle, middle of the night, the village erupts in the guttural barks and cries of the Yanomami tribesmen preparing for battle. I fly out of my hammock in time to see my guide duck into the hole that had been dug in the corner of our thatched hut— adrenaline over sanitation, he's just dropped into the jungle's version of indoor plumbing. Luckily, the cries and barks have nothing to do with us: we survive, though I can't say the same for the chieftain's wife, who they found with her lover all "midnight at the oasis"-style, tucked away in the jungle fauna. She was marched back to the village to meet her fate: they drove a large pole down onto the top of her head, caving in her skull.

[177] If I haven't said it before, I'll say it now: Tim is right, and I was very, very lucky.

history to knowing the third parties you've brought in or hired on (or are thinking of bringing in or hiring on). A situation—be it the deal you're making, the way your corner of the market is trending, the meeting you just walked into, or the IPR threat that your infringer's husband just hurled at you—is a living thing, evolving in real time, and the more you keep your eye on and track its moving parts (and what they all mean in context) the better you'll be able to assess and make the decisions that may mean the difference between a major payout and total disaster.

Information is power. This point and the next one are really more like subsets of the previous point. You want to understand as much as you can about what the hell is going on in the world of IP protection. At this point you should have a very good idea about how messy things are for the indie inventors right now, but things are constantly changing—new legislation is being proposed, new court decisions are being handed down—and all of these changes have a direct impact on *you*. Like it or not, your patent makes you a "stakeholder" in the USPTO and the IP world, now, and whatever happens in that world happens in your world as well. Any day now President Biden will select a new Director for the USPTO, and his or her policy decisions will affect your future for good or ill. Keep a finger on the pulse of current events by checking in with the websites I mentioned back in Part II. Knowing where you stand is the better part of self-defense and, like in any global hot zone, things can go from "calm and breezy" to "utter shitstorm" in the blink of an eye. Keep your head on a swivel and know your surroundings, even as they're shifting around you.

Trust. No man is an island, as they say, and the situations we encounter in the jungle and in the IP jungle have everything to do with the people we bring with, accept, or invite into our space. As the late, great Steve Irwin famously said, "Crocodiles are easy. They try to kill and eat you. People are harder. Sometimes they pretend to be your friend first." Or, as President Reagan said: "Doveryai, no proveryai; trust, but verify." The simple fact of the matter is that

185

—whether you're building a team to take you to the top of Everest, deep into the jungle, or through the maze of modern IP protection and market actualization—*due diligence* is the name of the game. People will jive and talk and dance to make you believe they can do it all, that they're the ones you should pick to run the ball down the field for you. Before you shake hands, sign the paper, or even hint at the possibility of a deal, ask yourself: 1) Is this person knowledgeable and experienced in this space? If so, what are his or her stats? 2) Is this person dedicated? Are they willing to give their all and go to the mat for me? 3) Does this person have integrity? Can I trust that they will do what they say they'll do?

And along with all of this, ask yourself: 4) Do I trust my own judgement? Do I have the ability to judge character? If not, who can I get to give me a second opinion? Because truth be told, whether you're a one-man or one-woman operation or the head of a loose affiliation of millionaires and billionaires, it's better not to make these decisions—or any decisions about the big stuff like your business—alone. If you're deciding whether or not to trust somebody, you'd be well served to bring in *somebody you already trust* and ask them to weigh in. It seems like obvious advice, maybe, but it's worth reiterating: even knowing better, I flew Han Solo into too many meetings with too many lawyers and potential partners—was carried along by the current of too many conversations and ended up agreeing to too many what-the-fuck-was-I-thinking? terms and deals—to just gloss over the point. Look long and hard before you leap. Weigh everything you know and can surmise about this person and then make the call. Entrusting any aspect of your IP journey to just anybody cannot be a snap decision. You don't want to fly to Vegas and get married to that person you just met on a blackout whim and you don't want to do the business equivalent, either. If this person really believes in you and your product and wants the relationship to work, they will meet you on your terms and timeline. If you tell them you need a week to think about it they will thank you for the opportunity, reiterate that they'd love to work with you, and tell you they're looking forward to hearing from you. People who pressure you to do things their way and on their timeline—who want you to agree and sign your name to terms *right now*—usually don't have your best interests at heart. To these sharks, you're just a little fish who swam within easy reach of their jaws.

Thinking about trust put me in mind of another old buddy of mine, and I reached out in the hope that I could pick his highly-trained and very well-informed brain for some insight on the matter. Tom Carter is a former member of the Diplomatic Corps (fondly called the "Dip Corps," practiced in the art of "bombs, bullets, and bad assery"); he's the kind of guy who used to get cryptic voicemails in the middle of the night saying things like, "The package has been delivered"... code that some missing high dignitary from some Third World country was "back in pocket" after 24 hairy hours of high speed chases, kidnappings, and coup attempts. My ex-Dip Corps pal has shifted gears since leaving the employ of the U.S. government: these days he makes his living as a corporate headhunter. For Tom's clients, a new hire may mean the difference between millions in profits and a disastrous quarter, all for the price of a kingly salary; at those stakes, there's little tolerance for error. Tom's advice? "Take the time, do the diligence, don't rush into anything, and never, *ever* sign a single sheet of paper or send an email that leads anyone to believe that you have a single opinion about anything even remotely related to 'commitment.' Be aloof. When composing an email or correspondence, *always* think about the endgame. If this email comes back in some evidentiary or discovery phase of a lawsuit, how's it going to look? What does it say, and what does it imply? Anything you write down can and will come back at you."

When it comes to partnering or hiring, Tom also passed along this killer strategy for learning all he needs to know: Start off with a one-on-one meeting with your prospective partner or employee. Something informal, maybe a breakfast. Then bring that person in for a group meeting with some of the other members of the team, or some people whose opinions you trust. Finally, bring them into a party or other large, informal gathering: something that involves cocktails.

187

"You'd be surprised what you can learn about your prospect by placing them in these different settings and seeing how they act," Tom told me.

This, from a guy who isn't surprised by anything.

Expect the hits. The patent process is a contact sport and, like in other contact sports, it's the hit you don't see coming that hits you the hardest. For me, that hit came from someone I deeply trusted, a lifelong friend, someone I adored and considered a brother. Someone who I'd made godfather to my oldest kid. Someone whose wife met me at the hospital when she learned that I was in labor and driving down from Westchester alone and in the middle of the night (my ex was in a meeting or on a conference call or doing something else oh-so important in another country). Massimo and I had been friends since back when a crushing forehand down the line was all that mattered to either one of us—back when a devastating serve mattered more than the fact that he was heir to the Ferragamo Empire, heir to the throne of everything faaaaabulous. Years later, when I was off and running with the CardShark, I presented the idea to him: I envisioned a version for the luxury folks who frequented his brand. I could see them sashaying into offices with their Ferragamo leather bags and leather shoes and leather belts... and a matching leather CardShark case. After all, well before orange, the smartphone was the new black. I was already (and still am) locked into various infringement fights with Massimo's peers in the luxury space: Gucci, Chanel, Louis Vuitton, Kate Spade, Yves Saint Laurent, Coach, Michael Kors, and Marc Jacobs were all ripping off my case, offering their own luxury combo wallet smartphone case. I told Massimo how tough it was to watch my design being sold at $400-$800 a pop—a tidy sum that I wasn't seeing any of. That's why the blurb he did for this book meant so much to me: reading it, I knew he understood what I was going through.

I never even considered having him sign an NDA. He was as close to family as it got. The fact that he passed on the idea did nothing to dampen our friendship, in my eyes. Not the right time, not the right season, not the right product. No big deal, these things happen. And then...

I was on a call with Eric Hurwitz. We were talking about the situation with some of the above-listed infringers. I said, "Well, at least I know that Ferragamo would never do that to me." There was a pause on the other end of the line, and then Eric asked me if I was sitting down. No, I wasn't, and I don't know that sitting down would have helped. I don't know that sitting down would have kept me from whirly-birding into free fall when I clicked the link that Eric sent over. There it was: the Ferragamo credit card wallet smartphone case, almost identical to the prototype I'd mocked up all those years before and shown to Massimo.

Like I said, it's the hit you don't see coming that hits you the hardest. I felt gut punched, blindsided. I texted Massimo a picture of the Ferragamo line of phone cases (there wasn't just one; there were many different designs) and simply asked him, "Why?"

I was willing to give him the benefit of the doubt. I was ready for him to tell me that he didn't know, that he was a busy guy running a $2 billion public company, that he didn't oversee all of the day-to-day operations and he didn't even know about the phone cases coming out of the accessory department. But none of that is what he said. Instead, he immediately replied that his lawyers had told him that their case did not infringe on my patent.

Which, as you know by now, is the first thing that every knowing infringer says.

Remember what I said about securing your helmet, keeping your head down and your mouth shut? I might add to that bit of advice that you stock up on Arnica, that homeopathic goop that soothes the knocks and bruises you'll inevitably receive along the way, even when you think you're doing that first part right. On top of that, maybe pick up my co-author Scott's book about Brazilian Jiu-Jitsu. Learn to redirect the power of the blows we are delivered, to bring your opponent in close so they can't get off a punch. And then...

Lawyer up. If you're setting off into an unknown jungle, you need to hire a guide. You need someone with a firsthand knowledge of the terrain, the people you'll encounter and their customs, the local flora and fauna, the "dos" and "don'ts." All of the tips in all of the Lonely Planet guides in the world don't hold a candle to this kind of boots-on-the-ground, baked-in, real-world knowhow.

All of which is to say—and, if you haven't realized it already, I'm sorry to be the one to tell you—you're going to need a lawyer. And here, maybe more than in any other facet of this undertaking, the rules stated above apply. Do your due diligence, decide what you need, and be unapologetic and unwavering in your decisions about who you do (and do not) want to work with.

In my experience, you're better off finding a legal team that is both patent savvy and user friendly in equal measure… even if that means going with a less overtly "prestigious" firm. The bigger the reputation—the more the firm is known for representing biggies like GE, Ford, IBM, Mobil, et cetera—the nicer their office location, the more they charge and the less interest they have in you, little fish. To a capable but smaller firm you'll be a big client and a high priority. Also, with these smaller firms (depending on the firm, its standing, and its policies), you may be able to negotiate a contingency deal: if the firm is established and willing, you could suggest a sliding scale: say, 20% for all fees collected for establishing a licensing deal out of an infringement case up to $100,000, and 25% on resolutions in your company's favor totaling over $100,000. You *can* find legal teams willing to work with you both on and for a fee structure that serves both your interests. They are out there.

Worst case scenario, go to your local law school and see if there are any graduates in patent law who can help you. Offer them a number on contingency that makes sense given the fact that they are not established and could use the experience. I'll bet you can come up with a deal that makes you both happy and benefits everyone involved. And I'll bet that that graduate's spry young mind is just as sharp as—if not sharper than—those of the liquid lunch lawyers from the swell Park Avenue firms.

But whatever way you go and no matter which path you choose or who you finally decide to sign on with, READ THE FINE PRINT on that agreement! The fine print is where they tuck in the clause about you selling them your first born. Good thing I had four kids... I still have a few left over. Because of this, crazy as it sounds, I'd even suggest that you consider *paying a second lawyer to read what the first lawyer wrote.* "What?" I hear you say. "With what money?" I know, I hear you. But you have to think of it as an investment. Remember what Tom said: "Always think about the endgame." If your business booms or if your deal with your legal representation goes south (like mine did), that fine print can mean the difference between walking away clean, done deal, and paying out thousands and thousands of dollars to sort it out on the back end. A few hours of a reasonably-priced lawyer's time to redline the contract is a small price to pay.

A fighter's mindset. This is something Tim focuses on a lot in his books and classes and seminars, and is another crucial point for those of you embarking on this path: your *mindset* must match your *context*. In terms of literal self-defense, think of it this way: it's all well and good to practice kicks and punches in a cardio kickboxing class, but it's something else entirely to direct one of those butt-toning kicks or shoulder-shredding punches at an assailant with the unmitigated intention to "do unto him" as he would have "done unto you." The *will* to fight and the *willingness* to do it are as important as the specific technical tools you put in your toolkit, and all the tools in the world won't save you if you don't take them out and use them.

What does that mean to you, the little indie inventor? Unfortunately, it means this: if you're heading down this road then you're in for a fight, one way or the other, and you'd better have your eyes wide open about 1) what's coming, and 2) what you're going to have to do to survive, thrive, and arrive at your final destination unscathed. I wish I could tell you that you'll be carried on the wings of butterflies, walking on rainbows from patent filing to brand launch to manufacturing and distribution and on through retail success... but you know I can't lie to you. It's going to be a war, and you need to know that going in. You need to be ready, mentally and emotionally. Filing a patent and launching a product or brand are exhilarating, make no mistake, but if you go in expecting the road to be smooth and the wind to be at your back the whole way then you're more likely to be thrown by the first little bump or detour that comes your way. Go in anticipating, to the best of your ability, those challenges and obstacles, and prepare yourself mentally and emotionally to face head-on anything else that comes your way, and you're going to be in much better shape when those obstacles arise.

However, none of this is to say that you don't fight smart, and part of fighting smart is knowing that "fighting back" and "going apeshit" are two very different things. As I said, I'm the last person you want to cut in front of in traffic; if you shove me I'm going to shove you back... even if you're a 500-pound gorilla. For me, then, knowing that I'm in for a fight and planning accordingly is all about making sure I have people around me to help me know when to hold my peace and when to swing for the fences: all about having people to hold me back when I'm about to throw myself into the jaws of a no-win situation. You may find that, for you, it's just the opposite: maybe, for you, understanding this principle means recognizing the fact that you don't like shoving back when you get shoved... and making sure that you bring someone onto your team who's ready, willing, and able to do the shoving when shoving is what's called for.

What does that mean? Most of the time it means—you guessed it—lawyers. Steven Wright said, "99% of lawyers give the rest a bad name," but oh, that 1%. I met some doozies my first few times around, but I'm extraordinarily lucky to have a team now that I know has my back. When I'm ready to fly off the handle about this or that infringer and their bullshit walletskins giving the CardShark a good hard shove right off the retail shelves, my team shoves back in a way that threads the needle between "lucrative payout" and "potential IPR." For you: see the points outlined above. Do your diligence, associate with people you trust, lawyer up, and be ready for the fight you're in for. Too many inventors spend years tinkering away in their garages only to finally step out into the light of day and get mugged of their ideas, their patents, their income, and all of the hopes and dreams those things carried with them. You want to do everything in your power to keep yourself from being one of them. As John Wayne famously said: "Don't pick a fight, but if you find yourself in one I suggest you make damn sure you win."

Take advantage. As Professor Liz alluded to, and despite the very real hurdles to effective IP protection for the indie inventor these days, in terms of the actual nuts and bolts of invention, there has arguably never been a better time to actually *be* an inventor. As Coleman Ban points out in his Edison Nation[178] piece "Why There is No Better Time Than Now to be an Inventor,"[179] advances in 3D printing have made it possible to "bring designs to life" in ways "that would never have been possible just a decade ago."[180] Digital collaboration technologies—FaceTime, real-time file sharing, and even YouTube and social media—make it not only possible but easy for like-minded inventors to work together on a project—even if they're on opposite sides of the planet. Online resources—like IPWatchdog.com and USInventor.org and others—offer new inventors an avalanche of information on patent and IP law, and "[w]eb sites like Kickstarter, companies like AKT IP Ventures and shows like Shark Tank are making it easier for anyone, anywhere to fund ideas that they believe in."[181] And, really, when we consider the low bar for entry to far-reaching advertising through social media and the distribution capabilities available to entrepreneurs through programs like Amazon Fulfillment, we see that the modern indie's landscape doesn't just contain obstacles: it also holds opportunities that would have been unfathomable to indie inventors working a generation ago.

---

[178] Edison Nation "connects innovators with companies to bring new products to market."

[179] "Why There is No Better Time Than Now to be an Inventor," Coleman Ban, www.Blog.EdisonNation.com, August 10, 2015. https://blog.edisonnation.com/2015/08/why-there-is-no-better-time-than-now-to-be-an-inventor/

[180] *Ibid.*

[181] *Ibid.*

In this vein, I'd like to introduce you to another comrade-in-arms I met on this crazy ride: Willy Ogorzaly—a.k.a. Willy O.—founder of LawBooth, who took time out from his very-impacted-wisdom-teeth extraction day to talk with me through his surgery-swollen chipmunk cheeks. Both LawBooth and Willy O. the entrepreneur are the product of the kind of fantastic resources up-and-comers can, should, and need to avail themselves of in the modern indie inventor sphere: Willy used co-working spaces while developing his idea, sought out and worked with incubator/accelerators (specifically the ones now cropping up left, right, and center in Boulder, Colorado), and entered (and won) student entrepreneurial competitions at school: all of which are great ways to build your idea, generate interest, and secure seed funding. As both the product of and an advocate for the new reality for startups, I wanted to hear Willy's thoughts on the current situation, and despite his recent medical intervention he was gracious enough to share them with me. He started off by telling me how, whereas Silicon Valley had long been the center of tech innovation in America, the trend was catching on and growing rapidly. Here's Willy:

Willy O.: *...Where before, if there were pockets of support for the startup community, places where the entrepreneurial spirit was really promoted, I would have to say that now the trend is spreading out across the country. Once it was only Silicon Valley. Now there's a growing entrepreneurial ecosystem.*

Kip: *What do you see as the key ingredients for this "entrepreneurial ecosystem," as you call it?*

Willy O.: *There are three: 1) a university presence, 2) access to capital, and 3) champion thought leaders. These three factors are why you're seeing these incubators and accelerators cropping up and thriving in places like Boulder, in Madison around the University of Wisconsin, in Boston around MIT, and so* *many more. 10 years ago there were maybe three or so of these capital accelerators, and now there are thousands. At the top you have Techstars, which is global, Y-Combinator, hubs like Capital Factory—which covers Texas—and Catalyze, which is CU Boulder's equity-free accelerator. The way it typically works with these accelerators is they basically take on 10-12 startups per quarter. Techstars, for example, provides capital, mentorship, workshops, and weekly check ups. The end goal is that during this three-month period the startup company's team is trained, they've built the business up with capital, and they have their pitch tight and honed. They're also expected to either raise funding during the program or prepare to raise upon graduating the program, which often culminates in a "demo day" to an audience of potential investors. The idea is that, by the end of the accelerator's three-month period of services, the startup is ready to go. They have all the tools and assets they need to really launch.*

Kip: *Can you tell me a little bit more about your startup, LawBooth?*

Willy O.: *LawBooth streamlines access to legal services by enabling prospective clients to schedule appointments with lawyers online. It's the first free, open marketplace in the legal industry. We created a platform where the entrepreneur has access to all of the legal resources he or she needs, and on the other side there's a healthy competition between lawyers regarding their costs and their capabilities. This transparency provides efficiency and an "invisible hand" that helps check the legal costs which, a lot of times, as you well know, can take entrepreneurs out at the knees. So many entrepreneurs find themselves held hostage by their own law firms. LawBooth is trying to solve that.*[182]

Big thanks to Willy O. for his input and his insights.

Keep shtum. I covered this idea a little bit in Part I, but it's more than

---

[182] Since the time of this interview, LawBooth has gone through a couple of changes. The company changed its name to JustLegal in 2017, and then to Got.Law when it was acquired later that same year by DigitalTown. So, when later on I tell you to check out Got.Law, don't be surprised.

worth repeating here: until you have your patent(s) signed, sealed, and in-hand, *shudduppayermouth*. Seriously, though: SHUT THE FUCK UP about your idea. I know it sounds paranoid, but you kindasortahafta BE PARANOID. Keep your idea on the DL. Even with things the way they are, even with the PTAB in place, getting your stake in the ground is the first step in taking a stand on, with, and for your IP, so until that's done keep shtum.[183] Loose lips sink ships, and if you're not careful they can very easily sink yours.

Fuhgeddaboudit. Take it from someone who's been there and who knows: once you've done all the legwork, paid all the lawyers, crossed all the t's and dotted all of the i's, and finally filed your patent application... *jahhsst fuhgeddaboudit*. Seriously. I wish I had better advice to give you, but I don't. The patent process is loooooong, even on the "expedited" track, and you're going to have plenty of time to sit and worry and wonder about whether someone beat you to the punch: plenty of time to lay awake wondering what if... what if...; plenty of time to stare, blinking, into the harsh blue light pouring out of your computer screen as you check for the 27,000,000th time to see if someone else's patent on your "good little idea" has slipped over the threshold between "pending" and "approved" since the last time you checked it nine to 12 seconds ago; plenty of time to check for the 27,000,000th time to see if this time that little hand grenade to all your hopes and dreams will come rocketing back at you from the depths of cyberspace... Trust me: *jahhsst fuhgeddaboudit*. Pretend it never happened. Let it go. Find zen. Worrying is bad for your health. Just take a deep breath and repeat after me: "If it's meant to be, it will be. Serenity now, serenity now..."

Act decisively. In fact, I might even go so far as to suggest that you try to *enjoy* the waiting—that you try to enjoy the peace and quiet while you've got it

---

[183] "Keep shtum (third-person singular simple present keeps shtum, present participle keeping shtum, simple past and past participle kept shtum): (intransitive, colloquial, idiomatic) Not tell anyone; especially, keep silent about something that may be sensitive or secret." https://en.wiktionary.org/wiki/keep_shtum

—because if and when your application comes back "approved" things are going to start coming at you fast and furiously, and you're going to want to be well rested and ready for it. When the situation calls for it, you're going to need to *act decisively*. For me, the first post-approval opportunity for decisive action—the first curveball that came barreling across the plate—was my lawyer asking me if I'd like to file an international patent application... and I whiffed hard. Huh? International what? My spiffy lawyer in his power tie was staring at me like *I* was supposed to know the things I thought I was paying him to know... and apparently my befuddled silence was enough to convince him that 1) explaining the situation to me would take too much effort, and 2) I couldn't afford it anyway. Which, hey, to be fair, I'd barely scraped together the money for the first filing, but it sure would have been nice to have someone explain all that went unsaid in the span of those two or three seconds—it would have been nice to have my fancypants lawyer lean in, place a caring hand on my shoulder, and assure me, "Hey, kiddo, don't worry. We've got you. We believe in this product a lot, and we're willing to go on contingency on this (new, previously unmentioned, $130,000) international patent filing fee"... especially given the chunk of change I'd already coughed up and forked over. I guess "ask me no questions and I'll tell you no lies; ask me no favors and I'll never disappoint you."

Again, see the previous points outlined above. In the IP jungle—as in the actual jungle—you live and die by your team, and you *must* surround yourself with people you trust to have your best interests at heart. By the time I even broached the topic of a contingency arrangement for the international patent the matter seemed—to him, at least—already settled. All I got was a smile, a shake of the head, and a "So sorry, but no, we don't go in on contingency, and therefore we can't cut you any deal on the international filing." Hopefully, if I've done my job right, this is the point where any person reading this book would know to walk, jog, run, sprint, swim, hitchhike... would know to do whatever they needed to do to *get the fuck out of Dodge*. Red means danger in any language, and the red flags were popping off hard and heavy... but what did I know? I thought I'd done what I needed to do. I thought I'd found people who knew what I didn't know: thought I'd found people who would steer me right.

All of which leads me to a related but important point—not quite a self-defense principle, but definitely a point of strategy. As you embark on this journey, ask yourself: "Where do I want to sell my great idea or product?" Think about your product. Is it something with universal appeal? Does it address a need or want that people share the world over? Is it something that the folks over in Timbuktu will want to use? Or China? Or Europe? If the answer is yes, then you're going to need to find a way to get that international patent. The bad news is that you need to be damn near independently wealthy to pull this off... or be really, *really* good at raising capital. The good news is that today there are AI systems that can scour the internet for infringers worldwide. You can subscribe to these sites and they will help monitor your patented idea based on your parameters and search words. Check out www.RedPoints.com for tips, advice, and to enroll in their protection. If you have that international patent then you can fight all those infringers out there. Me, I didn't understand the importance of what I was letting slip through my fingers, and I didn't have a legal team willing to spend the time to make me understand. If I'm repeating myself it's only because it's that important: *be very careful when you pick your team.* Good and helpful representation is out there, and—as per our conversation with Willy O.—there are more resources then ever at the indie inventor's disposal to help you find and connect with those people. Check out Got.Law[184] to start. You can tell them Kip sent you.

If you're not going to go international, then I suggest at the very least you **build your protective bubble**, i.e. you consider expanding your portfolio to include tangential sets of claims that border on but don't overlap with yours.

---

[184] Did you read the footnote at the end of my interview with Willy O.?

Like our buddy Adam Ullman said, patents are, by their very nature, narrow, and therefore somewhat easy to work around. If, for example, my patent only covered a smartphone case with a fully autonomous pocket, another company could design and manufacture a case with a slit allowing cards and cash to sit between the skin and the back of the phone itself without any infringement at all—no good for the little CardShark, with all its eggs (and cash) in one basket. Accordingly, after patent *numero uno* came back "approved," we set out to "build out my patent portfolio" and button up the whole world of combo wallet/smartphone cases. Patents two and three covered peripheral designs which, while they weren't necessarily anything we would ever choose to produce, kept other entities working in the space from skirting our primary design: they "covered our six"[185] so that we could focus on moving forward.

Haggle. Bringing a product from cocktail napkin to store shelf takes a lot of expertise in a lot of sectors and—unless you plan on learning how to do it all yourself—it means partnering with people and companies at nearly every step of the process. Some of the people you meet along the way will want to help you, some will want to consume you, but most will indifferently do whatever makes the most sense for their bottom line. I suggest you adopt this same mentality. With that in mind, it's time to close the loop on Carl, my plastics manufacturing mogul partner I introduced you to way back in Part I.

Carl was a great partner in a lot of ways, and at a time when I really needed one. Having someone like him in my corner—knowing that my idea had piqued the interest of a guy who got pitched a thousand ideas a day, and having the full weight and capabilities of his operation behind me—meant the world to me. I would say that the confidence and the sense of validation it gave me were

---

185 "'Got your six' and the related 'watch your six' come from service members telling each other that their rear is covered or that they need to watch out for an enemy attacking from behind." "15 common phrases civilians stole from the U.S. military," Logan Nye, BusinessInsider.com, April 19, 2021. https://www.businessinsider.com/15-common-phrases-civilians-stole-from-the-us-military-2015-6

priceless... but, of course, they weren't. They had a price, and it was one I didn't even question when Carl proposed it.

Carl was and is a very successful businessman, and it's not for no reason. He knows what he's doing, knows his numbers, and he knew how to offset the risk he was taking on with me and the CardShark. His percentage went across the board, and when the CardShark's business model was forced to pivot from manufacturing and selling to licensing, Carl kept getting paid. This was money I needed to keep the CardShark going, to live on, to pay my rent and put my kids through college—and each month Carl's percentage came out.

I should have figured a way out—should have cut a deal, paid Carl his fees for the molds and manufacturing fees, and moved on—but I didn't. I liked Carl. I liked having him on my team. He was (and is) smart and savvy, a tough businessman, and while we worked together he brought all of that to bat for me. Someone once said to me, "Kip, you need a pair of pants with you when you walk into these meetings." It would be nice if it was otherwise, but it wasn't (and isn't), and for the time that it worked I was glad that the pair of pants I brought with me into those meetings had Carl in them. He cut through bullshit better and faster than anyone I've ever seen. On top of that, though he always had the last word, it usually made me laugh.

When COVID landed on the world's doorstep, I finally made the call I'd been needing to make for years. I called Carl and I asked him if we could dissolve the original agreement. Licenses were all but drying up, and I needed every penny that was coming in just to keep on keepin' on. He agreed. He didn't have to. I'm grateful to him. I'm grateful for all he did for me. He was a good partner and an even better mentor, though some of the lessons he taught me were expensive ones. I wouldn't make the deal I made with him with another manufacturer today. I wouldn't just accept whatever terms they offered out of some misguided sense that I was the lucky one, here, and that they were doing me a favor. I know the value of what I bring to the table, now, and part of the reason I know it is because Carl showed it to me. He showed it to me with his support. Having someone like Carl see the value of me and my company and my idea helped me see it, too. To all you indies reading this, I wish you the kind of good fortune I had in finding a partner like Carl... but even more than that, I wish you the sense of value and self-worth I feel now. Remember that in business as in love as in everything else, "A good relationship is one where both parties feel like they're the lucky one."

All of this brings us neatly to our next set of principles: no matter the circumstances, no matter the deal, no matter the party, no matter the potential, always be willing to **stand your ground** and **just say "no."** In general, I try to live by the philosophy that I know what I know, and I know what I don't know. Accordingly, when I'm entering unfamiliar territory, I have no problem admitting my ignorance and/or limitations, and consulting the experts. And truthfully, I would advise you to adopt a similar philosophy: I've heard way too many stories of startups failing because of the ego of the inventor or core creator: too many stories of startups failing because the inventor or core creator kept only their own counsel... even when they were way out over their skis. As I think I've made clear, it's always a good idea to get with someone (or a team of someones) who knows the sector of the market in which you are creating and *listen to what they have to say*. However: this is not to say that you should question or doubt yourself or your own first-hand knowledge of your project. You may know what you don't know, but you also *know what you know*, and you should never let anyone tell you that they know better than you about what's right for you and your invention. Trust me on this. I paid plenty of retainers to plenty of self-proclaimed "experts in the field" when the person I should have been listening to all along was staring back at me from the mirror. No one else had invented the CardShark. No one else knew the need I was trying to serve, or the appeal of what I offered, better than I did. On top of that, I'd invented something beyond the scope of what the professional, so-called "expert" phone accessory designers and manufacturers had come up with. That certainly didn't mean that I had nothing left to learn in the space—

far from it—but it definitely put me a few rungs up the ladder from "utter noob." Me hiring these consultants was me forgetting all of this. It was me doubting my invention and myself. Take this advice from someone who's paid handsomely for the lesson: trust yourself. *You* got you here. There are people out there who know their stuff and who can help you get your concept down the field and through the goalposts, but never forget that they work for you, and not the other way around.

And, on a related note: just because someone tells you that they were instrumental in pushing the Snuggie or the Shamwow into the stratosphere of epic sales, that sure as shit doesn't mean they were. It's downright laughable to me how many people I've met who've told me that they "did the Shamwow." When someone comes along making big promises, telling you that they can help you carry the ball, they may be telling the truth—but you need *proof.* If they're really all they say they are—and if they're earnest about their intentions vis-a-vis you and your invention—then they won't mind backing up what they say with *data.* They won't mind showing you their stats, laying out what they have to offer. Remember: you're the patent holder. At the end of the day you're holding the ace, and that makes it forever *your* call whether to stay or walk. They have to convince you that *they're right for you*, not the other way around. Do your due diligence. Ask the big questions. What deals have they closed? What contacts do they have, and in what sectors? Do they know people in manufacturing, distribution, marketing? What do they bring to the table? What value do they add? Get their resume, call their references, Google the gigashit out of them, and don't you dare say you did when you didn't. It's not me you're bullshitting. The only ones you're hurting are you and your company.

And in general, before you spend any more of your hard-earned money hiring a consultant, think long and hard about what else that cash could be doing for you. The right consultant can bring a lot to the table, but depending on where you are in your journey you might be better served to spend that retainer putting real assets in place—like, say, that international patent that I should have (and wish I had) filed all those years ago. All the consulting in the world isn't bringing that one back.

Walk, talk, or run. No matter what you see in the movies—no matter how badass Keanu Reeves looks kicking the shit out of the bad guys who killed his dog—anyone versed in real-world self-defense will tell you that the best way to win a fight is not to get in one in the first place. Your chance of escaping unharmed plummets as soon as things go hands-on, meaning that it's always better to walk, talk, or run your ass out of whatever pickle you've found yourself in. Fighting should be the last resort of a cornered man, and so it is for you, little indie: avoid court like you're fucking allergic to it. You can't afford it, for one thing, and even if you think you can, you can't. Even if you want to, don't. Don't let your stubbornness drag you into it. Treat it like a pile of dog shit on the sidewalk and walk around it. If you find yourself in a situation like the one I've been in with the CardShark, the name of the game is to secure as many settlements as you can—ideally ones with ongoing license deals with the infringing company[186]—but if it's going to blow then let it blow over. You can't afford to get caught in the blast radius. See my L.A. story for reference. Did I walk away from a huge (potential) settlement? Yes. Was it the right decision? Absolutely. Going to war with that particular infringer would have meant an IPR, the likely loss of my patents, and, along with them, all of the royalties from the other licenses my team had managed to secure. That's the kind of fight where, just by fighting, you've already lost. When it's time to walk, walk and don't look back.

II. #CHESSNOTCHECKERS

This section is really a continuation of the last one, but is specifically geared toward all you "lady inventors" out there. First off: GET IT GIRL. You are a minority within a minority, and I fucking salute you. Second thing: You might want to sit down, because the news isn't great. Before I get into that, though, I

---

[186] See later in this Part, "MOVE OR DIE, or JUST KEEP SWIMMING," for more on this.

want to tell you a story. Don't worry, it isn't long. And hopefully, by the end, you'll understand why it matters.

A few years back I was out in Boulder, Colorado, getting my daughters all settled in for the start of the new school year at CU (go Buffs!). We were headed back to campus after our fourth trip to the Super Target, cruising along in the (rented) minivan I swore I'd never drive, and listening to the reoccurring *ding* of my eldest daughter (Kalypso)'s device alerting her that her online chess opponent had just made his move. It was a beautiful late summer evening, and we were riding along with the windows down, when all of a sudden the *ding*-punctuated quiet was shattered by Kalypso cursing in every language she knew how to curse in (and there are a few). Turns out some guy named Jacoby had been threatening her Queen, and now he was moving in for the *coup de grâce*. Within a few moves he had the Queen and was closing in on the King.

"Goddamnit," Kalypso seethed. "This stupid King. He does *nothing*. He can't move anywhere. The Queen's the only good piece on the board, the only one who can seriously cause damage, and now I'm screwed 'cuz Jacoby just ate her."

Listening to her go on, it occurred to me that I'd never really considered the power of the Queen. The King sticks close to his throne, forever limited to a lame, one-square shuffle, while the Queen enjoys the range and reach to swing down the board and knock any and everyone on their ass. And in life: it's the Queen who bears the children, who works to raise and nurture them, who's with them when they're frightened and when they're happy, who gets them to their soccer practices and piano recitals and Chinese lessons. It's the Queen who's there all the time while the King's away, orbiting a one-square radius around his throne far off in an office in the City. It's the Queen who's there waiting with dinner ready when he gets home late (again), waiting there as his Queen, his goddess of the grocery list, his Queen who's there to be whatever he needs her to be, his Queen who's there for years and years... until one morning she wakes up and she just can't do it anymore. She can't squeeze herself in and around his life anymore, can't be the Queen standing silently behind her King. She had dreams of her own, once, had hopes for her life and all it would contain, and this subservient suburban supporting role isn't on the list. And there's just no time. There's no more time to waste. Her life is passing her by one second at a time, and if she doesn't do something now then *this* is all she'll ever have or be. And the recoil off that one little realization, that screwed-down spring finally breaking loose, is enough to send blood and teeth and tears splintering across courtrooms and conference tables and therapists' offices well into the next decade. By the time she realizes what she's done—by the time she realizes that she's just blown up the world of everyone she cares about along with her own—it's already too late.

Divorce sucks, and it sucks worse if you don't know how to do it: if you married straight out of school and you never really "adulted" on your own. If you don't know how to navigate finances and investments. If you had no real clue about what it would be like to go from "being taken care of" to no job and no income. If all of your friends were couple friends, all of your connections were through him, and now all of a sudden you're cut off and cut loose. And here they are, these four beautiful kids, and they're looking to you to protect and shelter them, to *handle your shit* and be the parent they need you to be. And so you do. You hustle and you network. You hock the jewelry he gave you and you use the money that's left after you cover the rent on your new shitty apartment to run off resumes, to cover cab fare to job interviews. You work your core competencies, figure out where your added value will be rewarded. You crack open your 10,000th Red Bull and you brush the dirt off your shoulders and *you fucking get at it*.

And then, when you can, you hit the road on your bike, and you throttle back until life is just the speed and the wind and the blur of the world moving past you faster and faster. And maybe, if you're lucky, somewhere out there, far away from the flaming wreckage you've made of your old life, an idea comes out and meets you. And sure, maybe it's only a "good little idea," but it's yours, now. It's yours—that is—if you're willing to fight for it. And you are. You will. Because you're the fucking Queen.

I'm not special, and I'm not right. I'm just a chick who couldn't live on hold, couldn't live in orbit of someone else's existence, any more. I'm just a chick who wanted more, who looked inside and found the Queen who was waiting there: the Queen who runs the board, who moves with authority, who isn't afraid of her power. The Queen who embraces it all and moves forward. But here's the thing, my femme indie comrades-in-arms: that Queen is not me. She's there in all of us. She's in you, too: the badass Amazon warrior, who neither asks for nor requires anyone's permission to be. Who wants and strives without apology. Who makes her world over in her own image. It is you—us, *women*—who hold the cosmic power to create, to draw together the elements; it is *you who* are the true inventors, the true creators, the ones with the will and the mettle to bring forth something wholly new, wholly original, wholly sacred. You are composed of elements forged in the crucible of the stars, and you—yes, *you*—are fucking unstoppable.

Tenacity. Don't believe me that women are utter badasses? Then I've got another story for you. Barbara Carey is the brains and guts and hustle behind the $47-million Hairagami empire. You read that right: $47 million. Want to hear how she got started? She had an idea for an educational Halloween costume that she thought would sell well if she could get it onto Kmart's shelves. The problem was that, with Halloween only weeks away, Barbara was well behind the ball. On top of that, she didn't have a ton of resources at her disposal: she and her husband had nearly been bankrupted by an entrepreneurial venture gone bad. Nor did she have an inside line on Kmart's purchasers, or any contacts in manufacturing... To put it mildly, Barbara's dream was a very tall order. Still, Barbara wasn't going to let that sort of thing get in her way: she was going to get her costume into Kmart by Halloween, period.

With Halloween fast approaching she got on the road and headed east, charting a course from her home in San Francisco to Kmart headquarters in Troy, Michigan. She slept in her car, washed up in gas station bathrooms, and arrived in Troy to present the prototype of her "Who Am I?" disguise to the buyer... who shook his head.

"Why not?" Barbara wanted to know. Her product was good and her pitch was solid: the wholesale price and markup she was proposing should have made the order a no-brainer. On top of that, she looked great: she'd spent $4,000 of their last $6,000 on a red Donna Karan suit. ("Some people thought I was crazy to spend all that money on a suit," Barbara later said, "but it made me feel like a million dollars, and I was confident and looked my best when I met the Kmart executive.") She'd done everything right, so what was the problem?

"You're too early," the purchasing exec explained. "We're not doing next Halloween yet."

"Next?" Barbara said. "I'm not talking about next, I am talking about this Halloween."

"Impossible," the exec told her.

"But the price point is perfect, and I swear I can deliver."

Still: no dice.

Barbara wasn't finished, though. She didn't accept that. If this guy wouldn't make the deal, then she wanted to go up the chain. The next guy up said the same thing, though. Too late, too early, no deal. Fine: in that case, who's next? Barbara worked her way up until she was facing the big boss himself. She told him, "I can deliver in time for *this* Halloween."

And what could he really tell her? Standing behind her were a handful of embarrassed-looking execs with nothing to say. He told her, "OK. If you can deliver in time, you've got a deal."

Barbara Carey left Kmart headquarters with an order for 299,000 Halloween masks and an anticipated profit of more than $500,000. Her first stop? A payphone to call the manufacturer she'd met with on her eastward trip. He told her he needed to do a credit check on her to process the order. She told him, "Don't bother, it'll just show that I have no credit. You've got to trust me." She promised to pay him in 11 days if he could deliver the product in nine; he had her order ready in seven. She promised a $1,000 bonus to each of the 17 drivers who rushed the orders to their Kmart distribution centers on time, and then she went back to the bosses at Kmart to get her check. They told her, "We'll pay you within our 90-day pay cycle." She told them, "Like hell you will." She'd made a promise to her manufacturer and his drivers, and she meant to keep it. She waited until they cut her a check.

Barbara once said, "Why do I want to be one of Charlie's Angels, when I can be Charlie?" Exactly. Did I mention that she's worth $47 million? If you're looking to learn all Barbara has to teach you then I suggest you check out her book, *The Carey Formula: Your Ideas Are Worth Millions*, which she describes as a "roadmap to the American Dream." And if you ever find yourself scrambling, down to the wire and up against it, I suggest you do the following: take a moment to yourself, close your eyes, take a deep breath, quiet your mind, and then ask yourself: "What would Barbara Carey do?"

Fuck... off. You knew this section was coming. You knew because you understand the simple truth that—however we'd like it to be, however it should be—doing business as a woman means doing it in the context and company of men. The sad fact is that the business world can be a very ugly place for entrepreneurial women, and I'd be lying if I told you any different. Every step on the road offers pitfalls and hazards. The hunt for capital investment can easily turn into a tit-for-tat negotiation for the woman who doesn't know how to do the raising without teasing and taunting—and even the one who does.

Real talk, though: there's a subtle art to handling advances and managing expectations within the confines of the business world, where reputation and relationships count for so much. The fact is that "driven and smart and savvy" are very attractive things to be, and come with all the perks and handicaps of classic beauty. Case in point: one time, at an angel funding networking event, I was pitching to an ex-Wall Streeter who, while taking my business card, leaned in close to inspect the hue of my lip gloss.

"Extraordinary," he said. "I detect a slight sparkle in your lipstick."

Pop quiz, hot shot: What do you do in this situation? This human gland in a comb-over had enough capital reserves to fund me to the moon and back and not miss a meal; on the other hand... *ick.* I stopped mid-pitch, retrieved my business card from his manicured hand, and sashayed the fuck away. Not telling you what to do, just telling you what I did. Your mileage may vary. Just don't say I didn't warn you.

Here's another classic cautionary tale about a startup founded by two young, savvy, driven, and—importantly for our story—attractive business-women. Last I heard they were still under a gag order, but you know how word gets around. As with my buddy Denise Pierce, names have been obscured to protect the icky. Trust me when I tell you, though, that their startup idea was solid gold. It was so good, in fact, that it attracted the attention of a pair of fa-mous twin-brother investors (see how I haven't named any names?). It seems that one of the investors took a shine to one of the investees and, as she was pitching him hard, he dangled the necessary capital like bait in a trap.

"Great idea," he told her. "Here's the cash. No strings attached. Just sign here."

Remember what we said about reading the fine print? Turns out "no strings attached" took on a different meaning within the labyrinthian pathways of the twins' lawyers' legal scribblings—in ways that would have serious repercussions for our *femmetrepreneurs.* As a rich and powerful guy, the twin in question wasn't too keen on making do with life's little disappointments, and when our savvy smart businesswoman rebuffed his advances it set him on the hunt for a way to get even. When, down the road, the startup hit a rocky patch and need-ed to raise capital, an offer to buy the company outright looked like the thing that would save the day. The deal would have kept the company going—would have kept the lights on, the team together, the employees employed, and the end users supplied with a service they loved—but *no dice*. Somehow "no strings attached" had become "power to veto" in the contract, and the twins nixed the sale. With no other deals in the offing, the startup was forced to shut down. The twin in question cared less about losing his investment than he did about spiting the woman who'd rejected him. To anyone paying attention—and to any potential future investees who'd watched the whole thing go down—the message was clear: *screw or get screwed.*

Am I telling you this to scare you? Maybe. A little bit of fear has a way of clearing the sinuses and sharpening those critical thinking skills. But I'm mostly telling you so that you always keep at least the ghost of a backup plan percolating in the back of your mind, no matter how well it seems the first plan is working out. When your million-dollar investor who's crushin' on you hard and who fundamentally doesn't give a shit about your good or great idea gets his feelings hurt and wants to tantrum it out by pulling the plug, know your next move. Get up, brush your shoulders off, send your power suit to the cleaners, hold your head high... and get busy on plan B. They can knock you down, maybe, but they're going to find that keeping you down is a different matter altogether.

Be more kind. Let me end this section with one small but important piece of advice to all you *femmetrepreneurs* out there. If you should succeed—if you should happen to make it into the rarified air at or near the top of the entrepreneurial mountain—please: be kind to your sisters in arms. I will never forget walking into an all-women angel funding event several years back and being greeted by a firing squad of stares. These women—the other aspirants and the angels themselves—up-and-downed me like the editors at my second Sports Illustrated swimsuit issue callback. No skin off my back: I know a cattle call when I see one. Or maybe my pitch was less interesting to them than the question of whether my scarf was really Fendi or something faux. The point being: women can be as cruel as men, and there's enough working against us without us working against each other. Be nice to one another. Help in the boardrooms and angel rooms and investor rooms. If a young and savvy stunner in a business suit should grace your presence, rejoice that more women are shattering the glass ceiling and just be kind. Be helpful if you can, and supportive if you can't. Bitter and jealous women are just bitter and jealous women. Anyone with a creative idea and the tenacity to hold onto it is a badass, in my book, and deserves mad respect. And know that, if that describes you, I'll do whatever I can to help you. The Barbara Careys of the world are few and woefully far between, and the sad truth is that most of the few of us who start on this journey—even the ones who make all the right moves—don't make it to the end. The sad truth is that it's hard out here in the jungle, and a lot of worthy people and ideas fall by the wayside. Those of us who are still on the path have got to stick together.

—

They say that good judgement comes from experience, and experience comes from bad judgement. Like I said, I'm more guilty of the sins outlined above than anyone. Every bit of what you've just read has been learned on the sharp end. Still, I'd like to believe that my heart was in the right place: like to believe that I erred for all the right reasons. I wanted so badly to believe in people—young and old, female and male, any and all creed—that I as often as not failed to critically and objectively assess their assets. I put things on people that I wanted to believe they could and would achieve for me without always recognizing what was realistic, or even recognizing that what I wanted was not on par with the way things really work. I was bushwhacking my way through the underbrush with a lot of ideas about how things "should" or "ought to" work, and far fewer grounded ideas about the nature of the territory I was traversing. "Winging it" isn't a sound business model, and building the airplane

while you're already in mid-flight is certainly not the best way to go about it...
though sometimes it's the way these things go. I've had to learn a lot, and on
the fly. Like I said, my hope with this book is that, with the benefit of my experience, you'll have a better start and an easier path than I did. I wish you nothing but the greatest success and the best of luck.

III. MOVE OR DIE,
or JUST KEEP SWIMMING

*"The great white shark, the whale shark, and the mako shark don't have buc-
cal muscles at all. Instead, these sharks rely on obligate ram ventilation, a
way of breathing that requires sharks to swim with their mouths open. The
faster they swim, the more water is pushed through their gills. If they stop
swimming, they stop receiving oxygen. They move or die."*
—Meg Matthias, from the article "Do Sharks Really Die if They Stop Swimming?"

"Just keep swimming."
—Dory, *Finding Nemo*

So... "Where are they now?"
Or, in other words, "When last we left our hero..."
As I said, when I was starting out I had big dreams of manufacturing my
own CardShark walletskins, placing them with retailers, and growing the brand
that way. Unfortunately, as you now know, that's not how it worked out. Bat-
tling the infringers ate up all my capital, making it so I couldn't afford to manufacture; at the same time, I was a few years too late to the "sell your company
and make several million dollars" party. No investors wanted to step into the

shitstorm of patent litigation that had become the norm in the IP world. This all left me with licensing and the ongoing effort to convert infringers into licensees. I may be an NPE, but—as Monty Python says—society is to blame. We started with the first blatant infringer—that very big phone accessory company that thought they could steal my idea and roll me—and now we've got 15 of the top-tiered companies in this sector under our licensing belt.

There are advantages and disadvantages to licensing. An ongoing license means an ongoing royalty, and an ongoing royalty means that every quarter you see money coming in. Rack up as many of these licensees as you can, and maybe one day you'll be able to put enough aside to actually fund manufacturing and operations for your own production facility, and put your own version of your patented idea out into the marketplace. In the meantime, you—the creative genius behind your invention—get the freedom to keep doing whatever you want to do. Churn out new ideas for innovative products... or don't. No stress about starting a business, no headaches about manufacturing, promoting, and distributing your product. It's all on the licensee: all on them to hash the deals and forge the relationships. And bear in mind: the licensee is most likely not licensing your IP on a whim. Probably they're already operating in the space, already knowledgeable about the market they're launching your product into. A good licensee can turn your IP into substantial profit for all parties involved... and at zero risk to you, the patent holder. On top of that, these licensing agreements can last for years. On the other hand, it's all out of your hands. As Gene Luoma writes, "If you want to maintain control over your invention, the branding or packaging, and design, then you may not be a good candidate for licensing out the IP."[187] Someone else is dancing with your baby, now, and you may find that you need to make your peace with whatever direction they decide to take the idea.

Of course, proactive licensing is one thing: reactive licensing is something

---

[187] "What Is Licensing? Advantages and Disadvantages," Gene Luoma, ZipItClean.com. https://zipitclean.com/invention-news/licensing-advantages-and-disadvantages/

else entirely. As I've said, the licenses the CardShark has established so far have all been with former infringers, and that situation presents its own unique challenges. The tendency is to go in on the defensive. You've been ripped off, you're the victim, and here you are, about to sit down with your assailant. Still, the victim posture is a real soul-sucking bummer; also, justified or not, it's not a great place to negotiate from. It helps to feel like you're standing on a solid foundation, not struggling to get a foothold. For the CardShark, this sense of security has come from proving (and proving again, and proving again) that no such thing as the CardShark was ever created or patented before. Huge law firms representing big infringers have scoured the earth in search of any form of proof that my invention was anticipated or preempted by someone else's and so far, no luck. At this point it's been well established that the little underdog, the CardShark, got there first, staked its claim, and that our patent is legit; all of our licensing deals have rested on this foundation.

That… and the willingness to thread the needle between the cost of a license and the cost of an IPR. You know the deal. The truth is that the minute a big and deep-pocketed infringer decides to challenge my CardShark patents with an IPR, we're starting at zero. The fact of the existence of the licenses the CardShark has managed to secure, and the process we went through to prove our rights and their infringement, set no precedent. Am I tipping my own hand here? I don't think so. I'm only telling you what everybody already knows. IPR me and I'll be right back where I started, trying to prove that my patent is valid. I'll be a first-time defendant in the PTAB court. Doesn't quite seem fair, now does it? It doesn't seem fair that I should prove the validity of my idea to the USPTO, then to 15 infringers with the legal resources to fully investigate my claim, only to have to do it all again if an IPR comes down. That's the way it works, now. That's the way it is. But go on, motherfuckers. Bring it on. Do it. I double-dog dare you. It's all out and on the record, now. Anyone who wants to IPR me and strip me of my patents will do so in the glaring light of the obvious fact that the system is rotten and rigged from the get-go. You want to haul me into some PTAB court, and I promise you I'll put the whole system up on trial.

Still, there's a silver lining to this whole thing. I've squirreled away enough off a few royalty deals that I can now afford to manufacture my own kick-ass, carbon-fiber version of the CardShark for the iPhone 13, due out in time for Christmas 2021. And sure, I've been at this since the iPhone 3,[188] but who's counting? Carl and I made up 5,000 iPhone 3-fitting units and boxed them up all pretty; unfortunately, as you now know, it wasn't meant to be. Today, though, it's a brave new world once again: we have online selling capabilities that didn't exist just a few short years ago. Even Instagram and Facebook have buyer/seller interfaces that are load-and-go. On top of that, I've got a young engineer who's designed a new version of the CardShark that I really like (and who, as per the advice outlined in the previous section, I have vetted to the best of my ability). He's someone who will manage and oversee the CardShark's product launch and roll-out in a way that I couldn't, and carry it forward into the new era of production, online distribution, and marketing. Instagram influencers are being vetted as we speak; they really do seem to have the power to promote fans to click and buy.

The fact that nine years in I can finally think about manufacturing my own invention should tell you something about the current state of the patent system, and the new reality for patent holders. No matter how universal the need or how proven your idea is, you just can't count on funding from investors like in eras past. True, there are more crowd-funding platforms like Kickstarter every year, but the truth is that the burden falls more and more on the independent inventor to come up with his or her own resources. Fail to produce a Kickstarter response, and you can't bank on an angel to save you.

Still, I feel very hopeful about the new chapter the CardShark is starting. The new design—the way it's textured to slide easily in and out of back pockets—is like nothing any of the licensees or infringers have on the market. My plan is to distribute here in Chicago, and hire my base of homeless and at-risk military veterans to help me. Honestly, employing them is the better part of the reward: it sure beats any other minor (or major) victories the CardShark claims

---

188 Back when dinosaurs roamed the earth...

in the boardroom or the courtroom. Being able to help them—to give back even in this small, small way—is the best thing I could ever imagine doing.

IV. LET ME SEE YOUR WAR FACE

"If an idea, I reasoned, were really a valuable one,
there must be some way of realizing it."
—Elizabeth Blackwell, first woman to earn a medical degree in America.

"A single twig breaks, but the bundle of twigs is stronger."
— Tecumseh

Which brings us back to the question we asked at the beginning of this chapter: Where do we go now?

To look back at the origins of the U.S. patent system is to see enacted into law the regard and reverence our Founding Fathers held for innovation, invention, and inventors, and the central place they felt these occupied in the creation and cultivation of a new and thriving Republic. The America Invents Act fundamentally upended this central tenet, gutting the protections our Founding Fathers set out to establish. By the time our indie inventor friends climbed the steps of the United States Patent and Trademark Office to set their patents alight in 2017, more than 86%[189] of patents that had been IPR'd before the

---

[189] From the article "Federal Circuit Issues Second-Ever IPR Reversal, But Petitioner Still Wins," Justus Getter, DuaneMorris.com, November 24, 2015. https://blogs.duanemorris.com/ptab/2015/11/24/federal-circuit-second-ipr-reversal/ "Parties challenging patent validity in AIA proceedings therefore have an exceptional track record: successfully invalidating at least one challenged claim in 86% of Final Written Decisions at the PTAB."

PTAB were substantively invalidated. It's little wonder, then, that the turnout for the event was sparse: little wonder that employees of the USPTO couldn't even be bothered to step outside their own front door to witness the demonstration. It's little wonder that they didn't show up to defend their policies, didn't show up to explain why these protections—signed and sealed by their Office and supposedly backed by the full weight and authority of the U.S. government—were dropping like flies. Some attribute their absence to pressure from within the USPTO to downplay the entire situation, so maybe that's the reason. Maybe not. For myself, I'd like to believe it was something else. I'd like to believe that they stayed inside because they were too embarrassed to face the fallout from what they'd done. I'd like to believe that they stayed inside because they couldn't look those indies in the eye. I'd like to believe they were too ashamed.

Of course, really, it wasn't their decision. Cops don't make the laws, soldiers don't make foreign policy, and employees of the USPTO don't have any more latitude than any of these. For their actions to change, the policies themselves need to change. If you want to help—and I hope you do—you can follow and connect with the organizations and resources I outlined in Part II. Through them you can find out about new and ongoing efforts to petition, testify, fundraise, and otherwise communicate to policymakers in Washington the sad, boots-on-the-ground reality of the situation, and the need for substantive revision to the existing practices and policies. At the moment, one of the best things you can do is join in the movement you'll find described at www.USInventor.org/Resolution. By filling out the form you'll find there and adding your name to the list, you'll be joining with other indies and concerned citizens in a call on Congress to pass legislation to address the dire issues facing the patent system and patent holders. I would ask you to do this, and to encourage others to as well. The tech Goliaths and their lobbyists on the Hill already have the policymakers' ear; we need to join our voices together to shout as loud as they and their deep pockets do. We *must* come together in force in this battle for our rights. We need all hands on deck. We need *you*.[190]

What else can you do? You can think about your consumption. We can't get out of bed in the morning without interacting with hundreds and hundreds of patents, from your iPhone alarm to the motor in your electric toothbrush to the elements in the TV you click on to check the news. Think, for a moment, about the patent number stamped on each of those things, and the thousands and thousands more you will interact with throughout your day, and then imagine that—at the end of each of these numbers—is an inventor whose livelihood depends on your use of their product: an inventor who toiled away late into the night for years to create something, who gave his or her time and energy and talent and money to this thing that you get to casually consume and enjoy, this thing that makes your life just a little bit better or easier. Then think about how, by creating something wholly new, that inventor is creating jobs and opportunities that didn't exist before, providing opportunities to members of his or her community. Maybe that community is your community, too. And even if it's not, think about how that community's health is tied into your community's health: how growth and prosperity have a tendency to seep outward in a positive chain reaction. Support businesses spring up around central businesses; support industries spring up around hub industries. Maybe one day one of these inventors might provide an opportunity for you, or for your children. If you think about that, then maybe you'll pause the next time you go to buy the cheap, infringing knockoff on Amazon or some Whac-A-Mole website. Maybe you'll spend the extra $2 or $5 or $10 to buy the real thing: the licensed product or the one made by the patent holder themselves. You can think about it as a tiny investment in the kind of future you want. You can think about how you're helping someone like Gene Luoma or Linda Gomez or Adam

---

[190] Kip's note: FWIW, I plan on donating 10% of what I make from this book to U.S. Inventor to help them continue their work and in gratitude for all their help on this project. Big thanks, guys.

Ullman. Hell, if you want to, you can think about how you're helping someone like me. I know that I can speak for all the other indies out there when I tell you that we sincerely, sincerely appreciate it.

And for any of you who happen to be reading this from a position of power within our government—who happen to be a politician or policymaker with some say in the matter—first of all, let me say: Thank you very much for reading. Thank you for caring about these issues, for reading these stories, and for taking the time and making the effort on behalf of the indie inventors and the future of innovation they represent. I hope this book has made clear to you the problems that exist within the current iteration of the patent system, and has conveyed to you the dire urgency of the situation. I hope that you will consider Adam Ullman's point: that the patent system is integral to the composition of America itself; that the protections dictated by the Constitution are as central and sacred as those bestowed on roads and post offices; and that an attack on these protections is an attack on something the Founding Fathers enshrined as foundational and essential. I hope that you will use your position to voice these concerns within the Halls of Power, that you will communicate and amplify these stories, and that you will make your office the representative platform it was designed to be for the thousands and thousands of indie inventors whose cries have, for far too long, gone unheard and unheeded.

And for everyone else, indie or not, in this as in everything else: *vote*. Vote in people who understand the situation, who understand that decisions have come down that do not benefit the independent inventors and the small companies. Vote in people who see the issues and understand their priority. And vote out the people who don't. It may be bumped and bruised from the last go-around the horn—it may have had its fingers burned by misinformation campaigns by bad actors and foreign governments—but this is still a representative democracy, the last time I checked. Let your voice be heard.

And for all you indies out there: I'd tell you to keep pushing ahead, but you're going to do it anyway. As I said before, indies are a quirky lot. Statistically, some 87% of people are risk-averse: consistent, steady, 9-to-5ers, or some variation on the theme. Hell, even a fighter-jet pilot is kind of "risk-averse," when you really think about it. A mere 13% of the population are inventors or entrepreneurs, willing to take the road less traveled and give the dream a try. If you are one of these, I salute you. I wish you all the strength and courage and every color of the rainbow to bring your dream to technicolor life. I am here for any and all of you if I can help you in any way: if you want to reach me with questions, concerns, ideas, whatever, you can find me over at www.CardShark-Skin.com or at www.BloodintheWaterBook.com, where I'll respond from our collective email angrymob@bloodinthewaterbook.com. Let's connect. Let's join together and glow in the dark like the jellyfish fields that keep those great man-eating sharks away. We may be small, but together we can invent something new: a better, brighter future for all the indies working now and all the indies who haven't started yet: all the kids clicking away coding or taking apart their drones or putting together something I can't even imagine, because I'm not that singular genius with that vision and the drive to make it come true.

It is an uphill slog, but life is. I'm proud I've invented something that people use. Only a handful of people are driven enough to create and actually patent an idea. For you indies reading this, I hope you feel the same. I hope that reading this book, hearing some of these stories and interviews, has helped bolster and prepare you for the world we live in now, and inspired you to help fight to build a better one. It's a lot, I know, but I also know that you can do it. You are the indie inventor, one of the merry few who stick their courage to all that's best in us and take a flying leap into the unknown. It is for this reason that I have absolute, unshakeable faith in you, and in this country. I have faith that what's best in us will guide us through, if you keep showing us the way.

ACKNOWLEDGEMENTS

I owe some big, big thanks to the following people:

To Kalypso, Celeste, Sybilla, and Marco, my exotic-fruit-salad-named-kids, thanks for putting up with yer' Ma even when we'd ride the wave cresting it high only to be plunged back into the trough of the everyday. I know I rocked your boat and sometimes it was a bit stormy; through it all I loved you most and I pulled strength from your love, too. Heave-ho.

To my oldest friend, Maynard Brewer, the lobsterman I have the privilege of going sternman for, who once said to me, "Ain't it a wild river we live on." Why, yes it is. Thank you, Maynard, for always summing things up best.

To My Cousin Jennifer, a.k.a. Jeff-Jeff (she's going to kill me for calling her that), also a Maine lobsterman who could run with the best, who used to say, "Just hold on, you're in the barrel of it." She was my mom's favorite person because she knew how to quietly set about to just getting shit done. No drama, no chatter, she always just kept her head down... and quietly outpaced most of the guys around her. I have thought of—and drawn strength from—her example often throughout this wild ride.

To my Aunt June, an exceptional artist, who hails from some of the most rugged coastland in Maine (Monhegan Island to be exact); she's something special, and with a heart as big as the ocean. She's my hero and my inspiration.

To my Mom and Dad, their ashes now cast upon the waters up in Maine. I prefer to think of them as not so much gone as now simply everywhere. Through all these years they held my hand and my spirits up when I thought I had nothing left, thought I couldn't keep the CardShark going, thought I couldn't keep from being usurped and swallowed up. They listened to me moan, sob, cry, and shout... and then they helped me pick up the pieces and press on.

To my most trusted CardShark team, Eric Hurwitz and Karl Maersch, for looking out for me and protecting me in these shark-infested waters.

To my husband Mark: I've got no words to express how much your support and belief mean to me. When I first met Mark, I was the Chief Adventure Officer for a company called CompassionWorx, an app to help kids track community service and volunteerism. He was the first to say, "OK, how can I help you get this done?" With a shrug and a smile, he still laughs at my many adventuresome (if not certifiable) ideas... and then asks what he can do to help get it all done. For example: for my next act, I'm off to capture and share stories of veteran reintegration aboard the "Bullet & The Beast," a.k.a. the Airstream and the Ducati, with the Big Dawg (Geronimo) in tow. It is my hope that, by sharing these stories, I can help bring some much-needed attention to the struggles of our returning veterans, and help those who have sacrificed so much so that we can all "live the dream" (you can follow this project at www.radiocheck.tv). Mark is down for all of it. Ain't that grand?

Not least of all, to everyone who contributed their time, knowledge, guidance, encouragement, stories, and/or written material to this book: Eric Hurwitz, M. David Hoyle, Gene Quinn, Randy Landreneau, Josh Malone, Paul Morinville, Linda Gomez, Adam Ullman, Gene Luoma, Matteo Sabattini, Will Plut, Liz Stapp, Tim Wolf, Jeff Hardin, Patricia Duran, Agatha M. Cole, Tim Larkin,

Tom Carter, Willy Ogorzaly... and my biggest and sincerest apologies to anyone whose name I left off this list. This book would not and could not have happened without you.

Whether this book has resonated with you or not—whether it is destined to occupy a place on your bookshelf or destined to make a good doorstop on breezy nights—the simple fact is that it would not exist without the monumental effort of my editor-turned-co-author, Scott Burr. I don't know what it is about Scott that makes him a master of corralling the disparate and the complicated into a cohesive and comprehensible whole, all while untangling ideas from the gordian knots of my prose—whether it's the fact that, in his off hours, he's a novelist (check out *Bummed Out City*, highly recommended) and a Brazilian Jiu-Jitsu instructor (also check out *Worth Defending: How Gracie Jiu-Jitsu Saved My Life*, the memoir he co-authored with the first American student of BJJ); whatever it is, he's the architect who brought and held this house together and assured me that it could be done when I was equally sure it was all coming down. He had his hands full with this stubborn-ass student, and I am very grateful. I'm also very excited that we've got this one under our belt so that now we can tackle the rest of the absolutely batshit *craycray* stories that are still out there to tell...

And, last but certainly not least: to every humble inventor who ever had the inspiration to invent something and the perseverance to make it real. To this special breed, on behalf of the world, because the world too often overlooks where these ideas come from: Thank you.

*The authors at U.S. Inventor's **Decade of Stolen Dreams** Rally at the Midwest Regional U.S. Patent and Trademark Office in Detroit, Michigan, September 16, 2021.*

ABOUT THE AUTHORS

KIP AZZONI DOYLE is the creator of the CardShark, the original and still the best smartphone walletskin in the game. She is also a writer who attended Pomona College and then graduated from NYU before earning a master's degree in Screen Writing from the Tisch/Gallatin Division of NYU. As a freelance journalist she has written about the New York deep house club scene, kidnappings in Venezuela, the life of Pirate Queen Grace O'Malley, female Knights Templar, female racers (snow and asphalt), and the Yanomami tribe in the Amazon River Basin. She is the co-author, along with her former husband, Gian Luigi Longinotti-Buitoni, then head of Ferrari North America, of *Selling Dreams: How to Make Any Product Irresistible* (Simon & Schuster, 1999). Fusing her talent with mission—bringing awareness to the plight of returning veterans—Kip wrote and produced the PSA "The Other Note" and the short film "Until We Get Home" with Martin Sheen, both about suicide among our returning veteran population. She is currently working with a leading disruptor in the development space out in L.A. to produce "Radio Check," a TV series she wrote about soldiers, reintegration, and redemption. More information about the project is available online at www.LockLoadTV.org. Kip now lives in Chicago with her husband Mark Doyle and their 200-pound (he's lost 10 pounds on account of his enforced diet) St. Bernard, Geronimo. Connect with her online at www.BloodInTheWaterBook.com.

SCOTT BURR is a graduate of the creative writing program at the Colorado College. He is the author of the novels *Bummed Out City* and *We Will Rid the World of You*, the strength training manuals *Get a Grip* and *Suspend Your Disbelief*, and the martial arts, mindset, and health and fitness essay collection *Superhero Simplified*. He is the co-author of Richard Bresler's memoir of his over 40 years' involvement with Gracie Jiu-Jitsu and the Gracie family *Worth Defending: How Gracie Jiu-Jitsu Saved My Life* (www.WorthDefending-Book.com), which was an Amazon #1 New Release and an Amazon Top-10 Bestseller, and was the editor and designer for Robert Drysdale's bestselling book *Opening Closed Guard: The Origins of Jiu-Jitsu in Brazil: The Story Behind the Film*. He currently lives in Northeast Ohio. Connect with him online at www.ScottBurrAuthor.com.

Here it is: the "good little idea" that kicked me down the rabbit hole, and the reason you're holding this book in your hands right now. Nine years in and we're finally moving ahead with our own manufacturing. I can't wait. Check it out at www.CardSharkSkin.com.